# JavaScript Applications with Node.js, React, React Native and MongoDB:

Design, code, test, deploy and manage in Amazon AWS

Eric Bush

*"Give me six hours to chop down a tree and I will spend the first four sharpening the axe."*
*- Abraham Lincoln*

BLUE SKY PRODUCTIONS INC.

Copyright © 2018, Eric Bush

All rights reserved. No part of this book may be reproduced, stored in a retrieval system, or transmitted in any form or by any means, without the prior written permission of the author, except in the case of brief quotations embedded in evaluation articles or reviews as accompanied by proper references to the book.

Every effort has been made in the preparation of this book to ensure the accuracy of the information presented. However, the information contained in this book is sold without warranty, either expressed or implied. Neither the book's author, nor publisher, nor its dealers and distributors will be held liable for any damages caused or alleged, resulting directly by or indirectly by this book.

The views expressed herein are the sole responsibility of the author and do not necessarily represent those of any other person or organization.

Please send any comments to the author at jsdevstack@outlook.com.

Paperback ISBN-10: 0-9971966-6-1
Paperback ISBN-13: 978-0-9971966-6-5

e-book ISBN-10: 0-9971966-7-X
e-book ISBN-13: 978-0-9971966-7-2

# Contents

Preface ........................................................................................... 1

About the Author ........................................................................ 3

Introduction ................................................................................. 4

PART I: The Data Layer (MongoDB) ...................................... 13

    Chapter 1: Fundamentals ...................................................... 15

        1.1 Definition of the Data Layer .................................... 15

        1.2 Data layer design process ....................................... 17

        1.3 Introducing MongoDB ............................................ 19

        1.4 The MongoDB Collection ....................................... 22

        1.5 The MongoDB Document ....................................... 23

    Chapter 2: Data Modeling .................................................... 29

        2.1 Referencing or Embedding Data ............................. 29

        2.2 When to use Referencing ........................................ 32

        2.3 Reference Relationship Patterns ............................. 35

        2.4 A Hybrid Approach ................................................. 38

        2.5 Differentiating Document Types ............................. 39

        2.6 Running Out of Space in a Database ...................... 39

        2.7 Access Control ........................................................ 40

    Chapter 3: Querying for Documents ..................................... 41

        3.1 Query Criteria .......................................................... 45

        3.2 Projection criteria .................................................... 53

        3.3 Querying Polymorphic Documents in a Single Collection ..... 56

    Chapter 4: Updating Documents ........................................... 58

4.1 Update operators ............................................................................. 58

4.2 Array Update Operators ................................................................. 60

4.3 Transactions .................................................................................... 61

**Chapter 5: Managing Availability and Performance** ............................. 62

5.1 Indexing ........................................................................................... 62

5.2 Availability through Replication ................................................... 68

5.3 Sharding .......................................................................................... 70

**Chapter 6: NewsWatcher App Development** ......................................... 76

6.1 Create the Database and Collection ............................................. 77

6.2 Data Model Document Design ...................................................... 81

6.3 Trying Out Some Queries .............................................................. 85

6.4 Indexing Policy ............................................................................... 85

6.5 Moving On ....................................................................................... 87

**Chapter 7: DevOps for MongoDB** ........................................................... 88

7.1 Monitoring through the Atlas Management Portal ..................... 88

7.2 The Blame Game ............................................................................. 91

7.3 Backup and Recovery .................................................................... 91

**PART II: The Service Layer (Node.js)** ............................... 93

**Chapter 8: Fundamentals** ......................................................................... 95

8.1 Definition of the Service Layer ..................................................... 95

8.2 Introducing Node.js ........................................................................ 98

8.3 Basic Concepts of Programming Node ...................................... 101

8.4 Node.js Module Design ................................................................ 110

8.5 Useful Node Modules ................................................................... 113

**Chapter 9: Express** .................................................................................. 115

9.1 The Express Basics ...................................................................... 115

9.2 Express Request Routing ............................................................ 117

9.3 Express Middleware .................................................................................. 122

9.4 Express Request Object ........................................................................... 132

9.5 Express Response Object ......................................................................... 134

9.6 Template Response Sending .................................................................... 137

**Chapter 10: The MongoDB Module** ............................................................. 139

10.1 Basic CRUD Operations ......................................................................... 141

10.2 Aggregation Functionality ...................................................................... 145

10.3 What About an ODM/ORM? ................................................................... 146

10.4 Concurrency Problems ........................................................................... 147

**Chapter 11: Advanced Node Concepts** ....................................................... 151

11.1 How to Schedule Code to Run ................................................................ 151

11.2 How to Be RESTful ................................................................................. 152

11.3 How to Secure Access ............................................................................ 152

11.4 How to Mitigate Attacks .......................................................................... 155

11.5 Understanding Node Internals ................................................................ 160

11.6 How to Scale Node ................................................................................. 168

**Chapter 12: NewsWatcher App Development** ............................................. 177

12.1 Install the Necessary Tools ..................................................................... 177

12.2 Create an Express Application ............................................................... 177

12.3 Deploying to AWS Elastic Beanstalk ...................................................... 180

12.4 Basic Project Structure ........................................................................... 184

12.5 Where it All Starts (server.js) .................................................................. 185

12.6 A MongoDB Document to Hold News Stories ........................................ 190

12.7 A Central Place for Configuration (.env) ................................................. 191

12.8 HTTP/Rest Web Service API .................................................................. 192

12.9 Session Resource Routing (routes/session.js) ....................................... 194

12.10 Authorization Token Module (routes/authHelper.js) ............................. 197

12.11 User Resource Routing (routes/users.js) ............................................. 198

12.12 Home News Routing (routes/homeNews.js) ...................................... 207

12.13 Shared News Routing (routes/sharedNews.js) .................................. 208

12.14 Forked Node Process (app_FORK.js) ............................................... 212

12.15 Securing with HTTPS ....................................................................... 220

12.16 Deployment .................................................................................... 225

## Chapter 13: Testing the NewsWatcher Rest API ............................................ 227

13.1 Debugging During Testing ................................................................. 227

13.2 Tools to Make an HTTP/Rest Call ...................................................... 229

13.3 A Functional Test Suit with Mocha ................................................... 234

13.4 Performance and Load Testing ......................................................... 241

13.5 Running Lint .................................................................................... 242

## Chapter 14: DevOps Service Layer Tips ........................................................ 244

14.1 Console Logging ............................................................................... 244

14.2 CPU Profiling ................................................................................... 245

14.3 Memory Leak Detection ................................................................... 249

14.4 CI/CD .............................................................................................. 251

14.5 Monitoring and Alerting ................................................................... 256

# PART III: The Presentation Layer (React/HTML) ......... 261

## Chapter 15: Fundamentals ........................................................................ 263

15.1 Definition of the Presentation Layer ................................................. 263

15.2 Introducing React ............................................................................ 265

15.3 React with only an HTML file ........................................................... 266

15.4 Installation and App Creation .......................................................... 267

15.5 The Basics of React rendering with Components ............................. 274

15.6 Custom Components and Props ...................................................... 279

15.7 Components and State .................................................................... 281

15.8 Event Handlers .................................................................................................. 283

15.9 Component Containment .................................................................................. 285

15.10 HTML Forms ..................................................................................................... 286

15.11 Lifecycle of a Component ............................................................................... 287

15.12 Typechecking your Props ............................................................................... 287

15.13 Getting a reference to a DOM element ......................................................... 288

**Chapter 16: Further Topics** ................................................................................... 289

16.1 Using React Router ........................................................................................... 289

16.2 Using Bootstrap with React .............................................................................. 291

16.3 Making HTTP/Rest Requests ........................................................................... 293

16.4 State management with Redux ....................................................................... 294

**Chapter 17: NewsWatcher App Development with React** ................................. 302

17.1 Where it All Starts (src/index.js) ...................................................................... 303

17.2 The hub of everything (src/App.js) .................................................................. 304

17.3 Redux Reducers (src/reducers/index.js) ......................................................... 309

17.4 The Login Page (src/views/loginview.js) ........................................................ 312

17.5 Displaying the News (src/views/newsview.js and src/views/homenewsview.js) ........................................................................... 319

17.6 Shared News Page (src/views/sharednewsview.js) ...................................... 323

17.7 Profile Page (src/views/profileview.js) ........................................................... 328

17.8 Not Found Page (src/views/notfound.js) ........................................................ 334

**Chapter 18: UI Testing of NewsWatcher** ............................................................. 335

18.1 UI Testing with Selenium ................................................................................. 335

18.2 UI Testing with Enzyme ................................................................................... 338

18.3 Debugging UI Code Issues .............................................................................. 344

**Chapter 19: Server-Side Rendering** .................................................................... 347

19.1 NewsWatcher and SSR ................................................................................... 348

**Chapter 20: Native Mobile Application development with React Native..... 352**

    **20.1 React Native starter application** ................................................................. 353

    **20.2 Components** .............................................................................................. 356

    **20.3 Styling your application** ........................................................................... 358

    **20.4 Layout with flexbox** ................................................................................. 359

    **20.5 Screen navigation** .................................................................................... 362

    **20.6 Device capability access** ......................................................................... 363

    **20.7 Code changes to NewsWatcher** .............................................................. 363

    **20.8 Application Store Deployment** ................................................................ 374

  **CONCLUSION** ....................................................................................................... 378

**INDEX** ................................................................................................... 379

# Acknowledgements

This book covers some phenomenal technology frameworks. One of those is Node.js. I am in awe of Ryan Dahl for his having conceived of it in the first place and for his having built it on top of some great work by others. Thank you Ryan! I want to thank all those who have collaborated on the modules that make Node such a capable platform. This is open-source at its best.

I would also like to thank TJ Holowaychuk for his contribution to node through his work in creating the Express web framework. This really does a great job organizing and simplifying the code needed to implement a Rest Web API.

I want to thank the brilliant people behind all of the Amazon Web Services infrastructure. From those who envisioned it in the first place, to those who build it and also those who run it 24-hours a day. Every year it adds amazing platform services for all to take advantage of. This liberates application developers to code up their applications to provide customer value and benefit from many underlying services that do a lot of the heavy lifting.

Another great piece of technology is that of MongoDB. I want to thank those who develop and maintain MongoDB and also thank the Atlas team for their making it easy to host MongoDB in a PaaS environment.

I acknowledge all the effort that is ongoing to build, maintain and evolve React and React Native. The React framework from Facebook is a perfect fit in the JavaScript stack.

I express my love to my sweetheart Loradel. Thank you for finding me. I will cherish you forever.

# Preface

Many people will turn to the first few pages of a book to determine whether or not it will meet their needs. If you are just now making a decision about whether you want to invest the time and money necessary to build up your skills in the area of full-stack development, I will state that you are doing well to have found this book. Let me give you a brief marketing pitch about why you need to pursue investing in full-stack JavaScript skills.

Technology trends come and go, but what this book covers is really here to stay. The JavaScript technologies that this book covers have come together in a unified stack, all using the JavaScript programming language. With this book, you will not only learn about these technologies, but will also gain a longer lasting foundational understanding of what a three-tier architectural design is.

If you are even remotely interested in JavaScript development, I advise you to dedicate time to learn these technologies and it will pay off. You will have to study hard, but it will be worth it. There is a huge development community behind this and you will be in good company. Many top companies are finding success doing full-stack JavaScript development today.

**Level of skill required**
You will need to have some basic understanding of JavaScript to be effective in learning what this book contains. If you don't already know the basic concepts of the JavaScript language, you can use online content to learn them or you can buy a book to learn them.

I have written this book for all levels of experience. If you are a beginner and need to learn some of the basic concepts of a three-tier architecture, I have you covered. I always lead off with some material that covers the basic concepts that are involved in the architecture. I do this so that when you get to the actual code implementation, you will have this insight and are not just haphazardly writing code.

This book pulls together a lot of information that you would have a hard time finding. One of my goals was to give you knowledge that will help you get and keep a job. I have put myself in your shoes and have given you the information that you really need to gain traction.

If you are already familiar with the basics of the MERN stack, this book can get you to the next stage and actually lead you towards professional enterprise-level development. I make sure to dive deep into each topic and show real implementation and deployment details. In other words, this book is not a copy of the API documentation of each framework.

For the experts in MERN development out there, this book may save you time trying to figure out ways to test and secure your application.

Half the battle of learning any new technology is learning the add-ons and tools to get it all working. This is where this book will really come in handy. I don't just dump a lot of code on you, I take the time to explain the development process and what tools you can use.

**AWS and other knowledge**
This book utilizes AWS for its implementation. You get to learn about configuring and deploying to that environment. As mentioned, this book also contains substantial foundational instruction, such as details on security measures and information on how to perform testing.

**Choices**
This is certainly a great time to be a software developer. One of the things that makes it so great is the myriad of open-source projects available to leverage. There are frameworks that exist to help you with every layer of your architecture from the back-end services to the front-end GUI. There are literally hundreds of options for technologies to use in your development of a full-stack JavaScript application.

I will help you get started by narrowing things down to some of the essentials that are needed to get up and running. I will also cover more advanced topics to help you build quality enterprise applications.

I had to make some tough decisions on which frameworks and tools to utilize in this book and have done so based on the following criteria:

1. **It must be simple and quick to get value from.**
2. **It must have wide adoption, with committed support for growth and future relevance.**
3. **It must have broad platform reach.**

After mastering what is covered in this book, you can turn around and evangelize full-stack JavaScript development as a reality today. It is great already and will only get better.

Happy coding,
Eric Bush

*Note about a previous book I authored: Based on the above and many other factors, I made the switch from Angular to React for this book and changed the title. Many other changes and additions were also made.*

# About the Author

Eric Bush began his professional career programming in assembly language for embedded hardware microcontrollers for airplane flight systems. He then worked developing CAD software on what were known as workstation computers. When Windows came along, Eric started developing some of the first desktop applications for Windows. After that, he transitioned back to client-server applications for Windows Server. Finally, he made the move to enterprise cloud development in 2009 and is now focused on full-stack JavaScript technologies, using MongoDB, Node.js and React.

He has worked for some well-known companies including Walmart Labs, Microsoft, Tektronix, Mentor Graphics, Nike, Intel, and Boeing.

Eric worked as a full-stack developer for the worldwide retail giant Walmart. He worked on backend scalable JavaScript Node.js cloud infrastructure and also on front-end React code for the walmart.com website store details pages. The backend work serves up high-traffic product searches and cart capabilities used by millions of people each day. The work included many of the HTTP Rest endpoints in use. He was also involved in the DevOps work for Walmart enterprise practices of scalability, quality, security, reliability, availability and performance.

As a consulted at DocuSign in downtown Seattle he was in the role of an architect/developer and developed an internal enterprise application suite of tools to improve the efficiency of employees. This involved full-stack JavaScript with HTML/Angular that talked to a backend Node.js server that integrated with a MongoDB database.

As a consultant through Crosslake Technologies (crosslaketech.com) he puts his considerable breadth and depth of knowledge into helping companies implement modern PaaS cloud architectures. Crosslake helps companies envision, plan, architect, develop and execute engineering practices and architectures to satisfy business and customer needs. Crosslake works to transform and optimize software delivery. Crosslake also provides services for technical due diligence for mergers, acquisitions and investments.

You can email Eric at jsdevstack@outlook.com.

# Introduction

This book presents the technologies and best practices that span across all of the layers of a software development architecture. The core of this book focuses on specific JavaScript frameworks used to implement each architectural layer. Everything is based on MongoDB, Express, React, React Native and Node. I will walk you through the design and development of a sample application to teach you the best practices in architecting, coding, testing, securing, deploying and managing a RESTful Web Service and SPA (Single Page Application).

This introduction section will define several key terms, and introduce you to the sample application. This is important content that you will need before you dive deeper into the materials presented in the rest of the book.

## What is a development stack?

A development stack is the collection of languages and technologies used to construct the software application. These are the technologies you would piece together from bottom to top. This is different from what a software architecture is. An architecture is a pattern of components and layers for building software. This book utilizes a particularly popular set of technologies referred to as the MERN (MongoDB, Express.JS, React and Node.js) development stack. The architecture pattern followed is referred to as a SOA three-tier architecture.

MongoDB, Express.JS, React and Node.js are called platforms, or frameworks, because they take code, or some form of markup/configuration and execute it at a higher level of abstraction above the operating system level.

The MERN stack only specifies four technologies. However, in reality, there are many more technologies that come into play. Node.js as a framework really gets its capabilities from modular plug-ins that extend its basic capabilities. Platforms are usually extensible, meaning they are built to be extended in functionality by others in the community that provide modules. You will learn about many of the important extensions that can be utilized to extend the capabilities of Node.js, Express and React.

*Note*: *AWS does have database services such as DynamoDB, a NoSQL database technology. I have chosen to develop with MongoDB instead. This is because of its rich set of capabilities and its popularity. MongoDB is still hosted in AWS as a PaaS service, even though it is not provided directly by AWS.*

# Introduction

## The three-tier architecture

No matter what you are constructing, it is a good idea to modularize and build with layers so that you can more easily assemble everything. If you were building a house, you would not build it as one jumbled-up mess. Instead, you would first lay a foundation, and eventually, construct the attic and roof. Each of the needed modules would fit together: flooring, walls, heating, plumbing, electrical, etc. Software construction can be thought of in a similar way.

Architecting software in layers along with writing code as modules (also referred to as components) helps with testing, debugging, and incorporating future enhancements. The latest buzzword in the area of software construction is "microservices." You may also have heard of the older concept of a service-oriented architecture (SOA). These are patterns that encourage layers of separation.

One proven approach to developing a software application is to divide it up into three distinct layers. The three layers (or tiers) are named data, service, and presentation. You can also utilize patterns such as model view controller (MVC) within a three-tier architecture.

Definitions of the three tiers are listed below. An assessment is provided for where each MERN technology fits into those layers. While this list gives you the briefest of introductions to each of these, the three parts of this book go into the details of each layer.

- **Data Layer (MongoDB):**
  This layer is where you persist your data in a DBMS (Database Management System). Data can be stored, retrieved, updated, or deleted. MongoDB is categorized as a "document-based" database. This book will discuss a JavaScript API used to interact with MongoDB. A management portal called the Atlas portal and a tool named Compass will be used.

- **Service Layer (Node.js+Express.js):**
  This layer is where a web service API is exposed to contain your business logic and is the doorway to the data stored in the data layer. The service layer performs workflows that require more complicated computations and sequencing of operations. These might be HTTP endpoints or code that is scheduled to run periodically. Node.js was specifically designed for writing scalable server-side web applications. Express.js is used as a Node add-on module to simplify the building of a web service layer. Express has support for HTTP request routing and sending back of responses.

- **Presentation Layer (React/HTML and React Native):**
  This layer provides the capability to interact with and display data. This could be on a computer or mobile device of some type. React provides a way to code up components that render a UI by formulating the UI controls that drive the display.

*Note: In my opinion, the middle service layer choice of Node.js is something foundational to a JavaScript-based development stack and could not be swapped out. The other choices of MongoDB and React could be swapped out and replaced with some other similar technology choices. For example, since MongoDB is accessed through a Node module, you could easily find another solution using whatever database you prefer. React is the highest layer and could be replaced with any number of frameworks that allow you to make HTTP web requests to the Node service layer and do the data binding with either client or server-side rendering.*

## Using JavaScript everywhere

You understand from the book title, that 100% of the development done in this book is in JavaScript. There are considerable benefits with having one single language used in all three architecture layers. For one thing, the code will all look the same and be similarly refactored and maintained. You can also benefit from using the same testing framework across layers. Some of the libraries you download can even be used in both React and Node.js code.

JavaScript has huge momentum with lots of online resources and tons of available open-source code. You will soon become good at searching online before you start coding anything. Always search first, just in case there is a module out there that will do what you need.

*Note: You can decide to use TypeScript with your full-stack coding. The code for the usage of Node.js can still be the same as found in this book, as well as the usage of React and MongoDB. There may be minor differences, but those can be researched on the internet.*

*Note: JavaScript Object Notation (JSON) will be used as the data-interchange format. This is an excellent fit with a JavaScript application.*

## Using public cloud infrastructure

Every piece of code and all data storage should be hosted on public cloud infrastructure. This gives you characteristics like fast deployment, low cost, and elastic scalability. This book shows you how to construct an application that hosts everything in AWS.

As far as cloud hosting strategies, they can broadly be broadly split up into three categories. These are IaaS, PaaS, and SaaS. The "aaS" part of each acronym stands for "as a Service".

In **IaaS**, the "I" is for "Infrastructure" and means you can utilize the lowest level virtual machines (VMs or containers) and have complete control of what operating systems and software you deploy.

In **PaaS**, the "P" is for "Platform". This variation makes your life a bit easier as a developer. It still gives you the flexibility you need over machine deployment and scaling, but it frees you from the day-to-day maintenance of machines. You could go through the added work of using IaaS to install and manage your own Node.js service, or use a PaaS service instead that is much simpler to setup and operate. After all, you want to spend your time on application development, not on infrastructure provisioning and maintenance.

In **SaaS**, the "S" is for "Software" and indicates a complete offering such as Salesforce.com, Azure SharePoint, or Amazon WorkMail. Think of this as being applications that formerly would have been individually installed on your machine at work or home, but that now can be running on a machine in the cloud and accessed from anywhere.

*Note*: Since customers using a cloud service don't know when changes will happen to the service, you want to make sure they have uninterrupted access. For example, with Elastic Beanstalk as your hosting PaaS, your deployment updates can be "rolling" so that your service is never taken down. You will learn more about this. AWS also handles OS and hardware upgrades for you in a rolling manner that keeps your service up while that happens.

## Microsoft Visual Studio Code

If you are out in the real world trying to make a living by writing software, the reality is that you have to be able to combine many technologies at once and target many different platforms. Central to this is the code editor that you choose to use. Visual Studio Code (VS Code) is a great tool for editing your code. This is the tool I used to code and debug the sample application used in this book. You can certainly choose whatever code editor you like and are not required to use VS Code.

VS Code offers a rich editing environment as well as integrated features for source code control and debugging. VS Code also has the capability to launch tasks using tools like Gulp, without needing to jump to a command line prompt. These tools can be used to automate the build and testing steps that you run frequently.

Using VS Code, you can create a project, run it locally on your machine, and have access to IntelliSense, debugging, git commands and app publishing. VS Code can be installed on Mac OS X, and Linux, as well as on Windows.

If you are starting from scratch and don't have any development tools installed, go ahead and install the following:

- ✓ **Visual Studio Code**. Did I mention how cool this is?
- ✓ **Git**. This is used for source code control, but it can also be the means of deploying to AWS infrastructure through GitHub and other tools.
- ✓ **Node.js** from https://nodejs.org/. Get a stable version that is labeled LTS. This will also install NPM (a package manager for JavaScript projects) at the same time.

# The NewsWatcher sample application

The coding concepts in this book are illustrated using a sample application I called NewsWatcher. The sample application will help you understand how everything comes together across all three layers of the application development stack. I will now walk you through the capabilities and architecture of the sample application called NewsWatcher.

The basic capability of NewsWatcher allows a user to set up keywords to filter through news stories. The user can view, share and comment on the news stories. It will initially be developed as a website to run on a desktop or mobile device browser and then the UI will also be written to work as a native application for iOS and Android.

It is important to spend a short amount of time on the vision for what NewsWatcher was meant to be. I needed a vision statement and also some sketches of what it should look like. A rough plan for the development was laid out from there, including a prioritized backlog of features. I happen to prefer following the Kanban process to work through the work in a backlog. Here is the vision statement that I put together for NewsWatcher:

**Vision and Value proposition Statement**
*Users of NewsWatcher will get news personally served up to them. NewsWatcher will be their trusted advisor to alert them and save important information, without their needing to constantly scan endless feeds of news stories. Users can do things like adjust filters to get just the news they are interested in, see what their friends are looking at, and share comments about news stories.*

This vision statement is the longer-term goal of what to shoot for and would take many iterations to realize. From this vision statement, I was able to distill the vision down into a list of features that can be iterated over. Here is what I came up with for NewsWatcher. The list contains some requirements and I may have snuck in a bit of implementation detail also.

**Prioritized feature list that fulfills the vision:**
1. I can set up multiple news filters to hold news stories.
2. I can give my filter a title to identify it.
3. I can set up keywords for each filter with Boolean operations of AND and OR. For example, "Treatments AND insomnia", "Dogs OR Cats".
4. The filters are periodically scanned by the NewsWatcher backend service to search for news stories that match from a central pool of collected stories from a news service provider. Client-side processing will be freed up. Finding of stories proceeds even while devices are turned off or unable to connect to the internet. The server-side runs a periodic pull of stories from a news feed service.

# Introduction

5. News stories in folders can be scrolled through and clicked on to take me to the story content. The 20 most recent stories are shown for each filter.
6. Upon opening NewsWatcher, I see the home page of news stories.
7. I can click on and open a news folder to see the stories in it.
8. Stories in filter folders can be saved to an archive folder.
9. I can delete individual saved stories I select.
10. I can delete the filter itself with all of its stories.
11. My account settings are stored server-side, but cached on the device client-side also.
12. I can delete my NewsWatcher account.
13. There are globally shared news stories in which I can share a story to be seen by all NewsWatcher users. All other NewsWatcher users can see and comment on these stories. Stories will be kept for a week before being deleted. A maximum of thirty shared stories can exist; new ones will bump older ones off before the week is up. Any given user is limited to sharing five stories a week. There can be thirty comments per story kept. Offensive language in comments is blanked out.
14. I can view my news on whatever device I am logged in with.
15. The login expires and requires me to log in again periodically.
16. Two-factor authentication requires a text message code to be entered.
17. I can set actions on news folders to alert or email me when new entries are found.
18. I can set limits on when news alerts are sent to me. For example, wait at least ten minutes between alerts.
19. I can set times during the day when I don't want news alerts sent to me.
20. I can set NewsWatcher to download stories to my device when connected to Wi-Fi.
21. I can use a search entry box to do a search of all loaded stories.
22. I can designate the order I want to see the filters in.
23. I can email stories to others or share them via Twitter and Facebook.
24. iOS and Android phones have a Native app available in the app stores.

This book will only implement the core functionality of NewsWatcher in order to produce a Minimal Viable Product. Not everything will make it into the initial iteration, but it is nice to have a backlog ready to draw from.

**Wireframe prototype**

The following wireframe images were created in PowerPoint. PowerPoint has a great set of storyboard templates you can use to piece together a UI image and make realistic looking prototypes. There are also many nice websites you can use to sketch out a UI with.

To create a PowerPoint prototype image:
1. Open PowerPoint and create a blank slide.
2. Click the **Storyboarding** tab, then click **Storyboard Shapes** on the ribbon to open the **Storyboard Shapes** library.
3. Start creating your prototypes by dragging shapes from the **Storyboard Shapes** library to your slide.

# JavaScript Applications with Node.js, React, React Native and MongoDB

Figure 1 - NewsWatcher wireframes

**Note**: If you are using PowerPoint 2013, you need to have installed Visual Studio 2013 or later, or the Team Foundation Server Standalone Office Integration 2015 on the same machine to create and modify storyboards. PowerPoint 2016 is installed with everything you need for this feature to work.

## A peek at the result

Here are a few screenshots of what the actual Web UI ended up looking like on a mobile device. It doesn't look quite like the wireframes, as any good design will evolve.

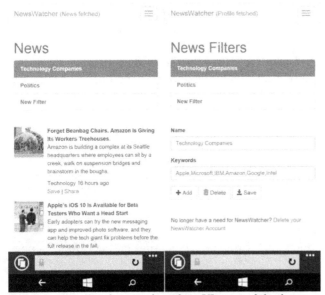

Figure 2 - NewsWatcher actual resulting UI on a mobile phone

# Introduction

You can try the completed sample browser app on your mobile device or desktop by going to https://www.newswatcher2rweb.com. Download the React Native app from Google Play.

All of the code for the sample application can be found on GitHub. You can download a ZIP file from https://github.com/eljamaki01/NewsWatcher2RWeb. The React Native code is found at https://github.com/eljamaki01/newswatcher2RN.

**A peek at the architecture and deployment topology**
The three-tier architecture of NewsWatcher is depicted in the following diagram. The arrows show what parts of the architecture call what other parts.

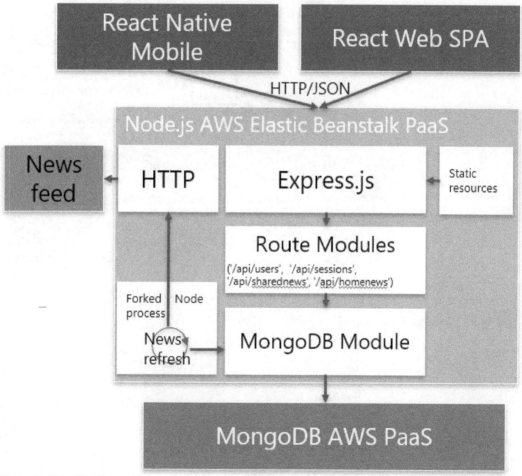

*Figure 3 - NewsWatcher architecture diagram*

The machine topology required to implement a three-tier application can vary widely and may even change over time. This is because the topology is really a function of the scaling that your application needs to achieve. You would certainly not start off with a topology that could scale to millions of users. It would be unnecessarily complicated to provision and maintain. You do only pay for what you need, however, some things would be in place that will still cost you money that you would otherwise not need.

The following diagram gives you a look at a starting point for an architecture topology. This topology would give you a fair amount of scaling to handle a large number of users.

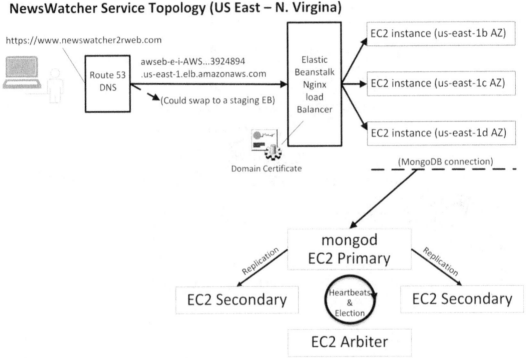

*Figure 4 - NewsWatcher service hosting topology*

# PART I: The Data Layer (MongoDB)

The first part of this book will now instruct you on what a data layer is and how to implement one using MongoDB. This will be hosted using DBaaS/PaaS in AWS using the MongoDB Inc. Atlas offering.

The very first step in creating a data layer involves the modeling of the types of data you will want to store. After that, you will learn what MongoDB is and how it can be used to implement a data model.

To finish out the first part of the book, you will actually construct a data layer that will support the needs of the NewsWatcher sample application. You will end up with a fully capable data layer that can be utilized by the service layer of the application architecture being built in the second part of this book.

In order to give full coverage to the topic, I will also cover what it means to manage the day-to-day operations of a MongoDB deployment.

# Chapter 1: Fundamentals

This chapter presents the concepts of a backend data storage system. I will show you what one is composed of and what capabilities are essential. After looking at backend storage systems in general, I will delve into the specifics of MongoDB and show how it is integrated into the NewsWatcher architecture to fulfill the backend data storage requirements.

*Note*: *This book is about full-stack development. While you will be shown how to get MongoDB set up in a PaaS environment and learn some basic development and monitoring, topics like what it takes to fully administer the service are out of scope for this book. Please see the MongoDB and the Atlas help resources for more information on administrative topics.*

## 1.1 Definition of the Data Layer

The data layer of a software application architecture provides for the persistent storage of information. Anything of importance can be stored there, such as customer account information, inventory, orders, audit logs, tax computation tables, and anything else that can conceivably be stored in electronic form.

The term database management system (DBMS) is used to describe the commercial offerings that implement a data layer at the lowest level. Each available DBMS has capabilities that differentiate it from the others. MongoDB from MongoDB, Inc. is one such DBMS system that is developed as an open-source project as well as being a commercial offering. See https://www.mongodb.com/ for their various offerings and capabilities. MongoDB Atlas (https://www.mongodb.com/cloud/atlas) is a commercial PaaS offering from MongoDB Inc. You will also find other companies, such as mLab (https://mlab.com/) that also host versions of MongoDB in a PaaS environment.

### DBMS capabilities

DBMS implementations can be placed into several general categories. One possible way to categorize them is as follows: relational, key-value, hierarchical, object-oriented, or document-based. Some DBMSs can also be said to be of the "NoSQL" type. Each category exists for a specific reason and you would want to look at your specific needs and choose the particular DBMS technology that fulfills your needs the best.

A DBMS will provide the physical storage medium where the data resides and should even withstand a power failure. In a cloud-hosted DBMS, data eventually makes its way to being stored on non-volatile cloud storage drives in a secure data center. Data storage might even be geo-replicated between distant data centers for redundancy and load-balancing.

# PART I: The Data Layer (MongoDB)

A DBMS will provide for the storage and access features for the creation, retrieval, updating, and deletion of data. The acronym "CRUD" is commonly used to refer to these operations. A DBMS will typically provide the following features:

- Data storage
- Transactions
- Attribution
- Auditing
- Authorization
- Data access
- Indexing

- Encryption
- Notification
- Programmability
- Schematization
- Security
- Transformation
- Validation

Even though the DBMS itself provides some type of programmatic access, there is often an additional layer on top that is referred to as a Data Access Layer (DAL). There are community written DALs as well as those that are commercially available, or you can even write your own. The DAL will create an abstraction layer that hides the complexities of the data storage technology and may even allow you to switch to a different backend DBMS without affecting the upper layers of your application.

A DAL can greatly simplify your access by creating an object structure that might not even exist in the actual data storage system itself. For example, some DBMSs don't provide schematization of data. If you need that, you can get it through a DAL. The following image shows all the sub-layers within the data layer:

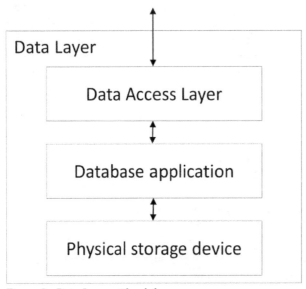

*Figure 5 - Data Layer with sub-layers*

Chapter 1: Fundamentals

# 1.2 Data layer design process

Before storing any data in your data layer, you need to create models for what that data should look like. Knowing what you would like to store is the first step to take. With that information determined, you can model what the structural form of the data will be. This will require some thoughtful design work to be able to organize your data in an efficient way. You will need to diagram out a model for your data to aid in understanding all the nuances of the record types with their properties and relationships.

User interfaces come and go, but backend data systems seem to live on. Often the data in a data layer outlives the applications that were written to expose it. It can sometimes be hard to go back and modify your data storage after you have started using it, so take your time and design it for future growth.

*Note: Even when the lifetime of the DBMS technology platform is finally reached, the data you have may remain valuable. If you decide to migrate to a newer DBMS technology, the data can be exported from your old DBMS and then be re-imported into your new DBMS. Most systems provide capabilities to export and import data in bulk. Some systems even provide an ongoing syncing capability for the transfer and transformation of the data.*

## Data layer planning

Make sure to utilize experts along the way before rolling anything out into a production environment. The following questions are crucial as you work through the initial data layer design. The answers to these questions should be carefully determined.

Data layer planning questionnaire:
- Is the data shared? If so, how is it shared between customers, applications, and processes?
- Is multi-tenant storage ok? What isolation of data is necessary?
- What are your data security and privacy requirements?
- Will you be storing data that is classified as high business impact (HBI) or storing any personally identifiable information (PII)? If the data is compromised, what are the legal ramifications?
- Do you need data access roles to control access to the data?
- Are there any periodic processing jobs that run that will access the data?
- Will parallel access to a single record cause concurrency issues? Is there some type of locking or serialization of access required? Would request queuing or optimistic concurrency control (OCC) suffice?
- Do you require transactional capabilities?

# PART I: The Data Layer (MongoDB)

- Can writes be asynchronous or do you need immediate synchronous acknowledgment on a write?
- Is a lag time between write and availability of that data for a read operation acceptable?
- What is the service level agreement (SLA) requirement for each CRUD operation?
- What are the access volume and rates per minute? Will there be bursts of activity or is the activity evenly distributed across each day?
- What is the size of the data? How many records and how large will they become?
- How many users will access the data? What will be the needed data capacity per customer?
- Do you anticipate that the structure of your data will change?
- Do you need to keep a record of each data access, such as keeping an audit trail of accesses and changes?
- Will you be running data mining and business intelligence analysis over the data?
- Is there a strict data schema that needs to be validated against?
- What are the record types, contents and relationships?
- What system dependencies are involved? How is the data transferred and how is it combined and verified?
- Are you storing media data that is typically in binary file form such as photos, movies etc.?

Chapter 1: Fundamentals

# 1.3 Introducing MongoDB

MongoDB is classified as a NoSQL database. This means that it is non-relational and non-schematized. There is no central schema catalog or data record structure definition required. That can exist, however, it is not required. Database records can basically be free-form. Don't get too worried about the unstructured nature of MongoDB records, you will learn how to set up a data model and how to validate data before it is stored in your data layer.

MongoDB is also classified as being document-based. It is a document-based database because it conceptually stores JavaScript Object Notation (JSON) documents. I say conceptually, because it does not actually store JSON documents directly, but instead stores an internal binary representation. Documents are stored in Binary JSON (BSON) form.

Document-based storage with MongoDB really hits a sweet spot for modern application needs. It gives you the best of scaling and performance. Because it can be purchased as a PaaS-hosted service that runs on AWS infrastructure, it is an easy to manage environment. You can, of course, install and run it yourself.

*Note: If you look at the database offerings of AWS, you will find several choices for data storage in the cloud, including DynamoDB, RDS/Aurora, Redshift and S3. Each offering exists to fulfill different requirements, and each has its own advantages. MongoDB is offered through MongoDB Inc.*

Companys such as mLab or MongoDB Inc. offer PaaS solutions. The Atlas product from MongoDB Inc. can be installed as a service and run in cloud-hosted services such as AWS. Through the Atlas management portal, you sign up to use MongoDB and then your database cluster (replica set or sharded cluster) is deployed to cloud infrastructure on your behalf.

*Note: MongoDB itself is built on top of another open source component called WiredTiger that is a data storage engine. This is something that is not necessary to know about. MongoDB is all that you will interact with directly.*

## Benefits of MongoDB with Atlas PaaS

Figure 5 in the prior section showed an illustration of the data layer. This included three different sub-layers. The great news is that you get all three with MongoDB and its API access. Here are some compelling features to consider when evaluating the benefits of MongoDB, especially through the Atlas offering that uses cloud infrastructure:

- **Elastic storage capacity:** You simply dial up and down your storage capacity by moving to a higher performance plan or by adding more capacity. You can delete

infrastructure that is no longer needed and then not be charged for it anymore. You can configure your resources through code or through a management portal. There is practically unlimited growth for "pay-as-you-grow" storage capacity.
- **Elastic performance:** You simply move your performance up or down by selecting a different plan. To achieve higher throughput, you can pay for the highest performance tier and get a sharded cluster and the highest IOPS.
- **PaaS:** MongoDB on AWS through Atlas is a PaaS offering. All the machine management of software and hardware upgrades are handled for you.
- **APM:** Capabilities for monitoring, alerting, scale management and backups keep you in control.
- **Features:** Besides creating, reading, updating and deleting documents, there are many features to use under the right circumstance. Features such as views, indexing, replication, change streams, sharding, aggregation, capped collections, TTL indexes, access control, encryption, ACID transactions and much more.
- **Auto-Replication:** Data is automatically stored in redundant copies of the database. The replications are there for your safety, to ensure availability through failover. You can also set up automatic backups to happen.
- **Security:** MongoDB has an audit log to track all database operations. For added security, you can have an SSL connection to your database. MongoDB also has an encrypted storage engine available to protect data.

The Atlas MongoDB offering on AWS is ready to meet all your business and customer needs. It has a great feature set and is increasing in capabilities all the time.

## Try it out

If you really wanted to, you could download MongoDB and run it on your local machine. You could also do the work to set up a cloud-hosted VM or use a Docker container with MongoDB. My preference is to use MongoDB as a cloud-hosted PaaS solution.

MongoDB Inc. makes it easy to sign up for a plan that then hosts MongoDB on your choice of Amazon AWS, Microsoft Azure, or Google Cloud Services. This is the approach I have used for this book. You would have to do your own research if you want to depart from that. The rest of this book is written with this approach in mind.

It takes just a few clicks to and be up and running with MongoDB as a hosted service in AWS. You don't need to worry about the daily details of managing the software updates and machine hardware maintenance to keep it up and running. There is no need to waste any time dealing with hardware failures, such as the replacement of drives or network cards.

You enjoy the benefit of having machine replication, scaling, load-balancing, failover, and backup. With the scaling of MongoDB, you only pay for the storage and performance you desire. You pay for what you use and can easily scale down when you no longer need as much data storage or performance.

You are free to concentrate on the aspects of your application that deliver value to your customer and increase ROI for your company. There are many benefits with PaaS cloud infrastructure.

## Pay for performance

As previously mentioned, you only pay for what you need. As with any cloud infrastructure, if you have an immediate need, you can pay for higher performance machines to run MongoDB. Thankfully, there is a free option through Atlas for your initial investigations.

You can also pay for your higher scaling capability by purchasing a plan that comes with a higher storage allotment and is configured with more machines for high-availability and auto-failover. If you need to scale up to more storage, you can keep adding database clusters as needed and spread your data across more machines with sharding.

You also have the flexibility to pay for more than one plan at the same time and put data on the more expensive plan that needs the highest throughput, while other data can be on the less expensive plan of storage. You can change things like the instance size, replication factor and sharding scale as needed at any time.

In some cases, a configuration change will still require downtime, such as a change in the instance size. This downtime would happen while the primary server is being migrated. But this time should be under a minute. See the Atlas documentation for more details.

## MongoDB structure

With your Atlas account, you can create one or more database clusters. These clusters then have databases that serve as the containers for what are called collections. Databases hold collections and they also hold your indexes and user accounts. Collections, however, are what contain your data in the form of BSON documents.

Each database can have a set of users with specific permissions. You can control the users of your database and give individuals read access and also write access. This does not mean that everyone that connects to MongoDB through the NewsWatcher app needs their own user account. General programmatic access happens through your service layer. Specific data access is controlled through a middle-tier login mechanism. That information will be covered later in this book.

*Note: You can create user accounts at the cluster level that work across all databases created on that cluster. An account is created when you first create the cluster, for admin privileges.*

PART I: The Data Layer (MongoDB)

The following illustration is an overall visual representation that shows the different resources that are part of the MongoDB managed platform service and how they relate to each other. The important thing to understand is that you can have multiple databases, with each database having multiple collections in it.

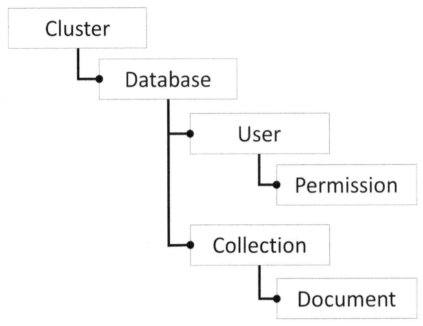

*Figure 6 - MongoDB resources*

# 1.4 The MongoDB Collection

A collection is a container for storing documents and is the main resource with which interactions happen programmatically. You can write JavaScript code in your Node.js service layer for operating on the data in the collection.

Collections are a convenient way to separate out data in a MongoDB database. They collectively share whatever the limit of total storage is for the plan you are paying for.

To write data into a collection, you will want to use one of the APIs available to do that. You can also write data using the MongoDB command line tool, or the Compass UI application that can be installed on your machine. You can add any documents by hand. Most likely, you will want to use one of the code language APIs available.

Chapter 1: Fundamentals

# 1.5 The MongoDB Document

Every database ever invented is fundamentally designed to store data as a set of records. Each individual record contains information that is useful for later retrieval. For example, in a relational database, the record is in the form of a row in a table. For example, you might have a table of customers where each row represents a single customer with their name, age, and email as part of their record. In AWS S3, a record takes the form of data with its descriptive metadata.

In MongoDB, a single record is represented by a document in a collection. The following image illustrates how you can have multiple collections, with each collection containing multiple documents. Each rectangle is a document below:

**Collection One**

```
{
  "email": "eb@hotmail.com",
  "filters": [
    {
      "filter_name": "Pets",
      "keywords": ["dog", "cat"],
      "newStories": [
        {
          "title": "Life on Mars!",
          "link": "http://...",
          "img:": http://...,
          "src_date": "CNN 1 hr ago"
        },...
      ]
    },...
  ],
  "settings": {"enableWIFI": true,
  "enableAlerts": false}
}
```

```
{
  "email": "xyz@hotmail.com",
  "filters": [
    {
      "filter_name": "Politics",
      "keywords": ["US", "election"],
      "newStories": [
        {
          "title": "Election results!",
          "link": "http://...",
          "img:": http://...,
          "src_date": "CNN 1 hr ago"
        },...
      ]
    },...
  ],
  "settings": {"enableWIFI": true,
  "enableAlerts": true}
}
```

● ● ●

**Collection Two**

```
{
  "RentalType": "Home",
  "Address": "123 Main street",
  "Size": "3 Bedroom",
  "Price": 2000
}
```

```
{
  "RentalType": "Condo",
  "Address": "789 Center street",
  "Size": "2 Bedroom",
  "Price": 1200
}
```

● ● ●

*Figure 7 - Example MongoDB collections with documents*

Documents in this example, are represented as JSON with top-level curly braces enclosing each document. This is how they are shown textually, even though they really do not appear that way in the database itself.

This simple concept of storing documents in a document-based database has many advantages. JSON is easy to formulate and consume in code. Another benefit is that it can contain complex hierarchical data in embedded structures. It also supports arrays.

## Datatypes in a document

The primitive data types supported in a MongoDB JSON document are the same ones that are available in JSON. These are:

- Array
- Boolean
- Null
- Number
- Object
- String

In addition, there is an extended syntax in JSON that can take the augmented BSON datatypes and preserve them. Some of the added data types of BSON are Date, Binary data, 64-bit integer and many others. You will eventually be accessing MongoDB through code and will be using JavaScript objects with their supported datatypes.

I will now give you a few tips for storing media data and currency values, as there are some interesting techniques to utilize.

One of these issues is with the storage of binary data, especially media data such as photos and movies. Media data cannot realistically be inserted into a JSON file. You could use the BSON binary data type, but the main problem you would encounter would be that the maximum BSON document size is 16 megabytes. A movie would simply not fit into one single document. It is best to store binary data in a separate storage system, such as AWS S3, and reference it from a MongoDB document.

Another issue is the representation of currency values, such as US dollars and cents. The problem comes when you attempt to perform floating point precision storage and arithmetic operations, such as might be required for investment calculations. If Node.js were to calculate and print 0.1 + 0.02, the result would be 0.12000000000000001. You are not even safe in using the MongoDB 64-bit floating point datatype. This is due to the rounding errors inherent in CPUs because of how they represent floating point numbers.

One solution is to multiply all your monetary values by a constant scaling factor. For example, if you were going to store the value of $1.99, don't store 1.99, but instead multiply that by 1000 and store an integer value of 1990. Then you can perform math on those integer values and convert back for display when needed.

# The JSON document

One distinguishing characteristic of a document-based database is that it can store complex hierarchical objects without any predefined schema. This means that every document in a collection could have a different structure. This could be considered both an advantage and a disadvantage.

There are many advantages to having a schema-less database. The obvious one is that you don't need to specify the JSON format in advance. This means that there is no need to specify datatypes or have restrictions on them. It is simply a matter of creating the JSON with the name/value pairs that you desire and then inserting that into a collection. Here is a simple JSON document:

```
{
    "_id": ObjectID("59612a3dc17c5416d0a33041"),
    "myNumber": 99,
    "myString": "Hi there",
    "myBool": true
}
```

*Note: I refer to the name/value pairs in the JSON Document as "properties", since its syntax is so close to the syntax of the property in a JavaScript object literal. The MongoDB online documentation, however, uses the term field/value for the pairs.*

If you open of the MongoDB Inc. Compass app and look at some documents, you can see each in the UI with all their properties to explore. In this chapter, documents will be presented in the textual JSON form since you can visualize them better that way, and if needed, import them into a collection in that form.

Once you get to the chapter on Node.js and are using the API to programmatically interact with MongoDB, you will see JavaScript code with the object literal syntax used. Just be aware of that switch. As an example, the object literal syntax equivalent of the prior JSON example would be as follows (MongoDB will provide the id property upon insertion):

```
var mydoc = {
   myNumber: 99,
   myString: "Hi there",
   myBool: true
};
```

*Note: There is one restriction on JSON documents to make them work in MongoDB. The restriction is that the property names in a single document must not be duplicated. This makes sense because, if you were to query and ask for a given property and it existed twice, that would be a little confusing. The core MongoDB storage system that stores BSON does not make this restriction, but the Node.js module API that accesses it requires it.*

PART I: The Data Layer (MongoDB)

# An example

Let's pretend you are running a business that sells books over the internet. The following document represents what you might want to use in a collection that holds customers of your online bookstore. Each customer document would hold information associated with that customer. You would want to store information about what books each customer had purchased. The book order data for the customer is embedded in their document. You would also want to store their personal contact information. All of this can be placed in one single document as follows:

```
// Customer Document
{
   "_id": "77",
   "name": "Joe Schmoe",
   "age": 27,
   "email": "js@live.com",
   "address": {
      "street": "21 Main Street",
      "city": "Emerald City",
      "state": "KS",
      "postalCode": "10021-3100"
   },
   "booksPurchased": [
      {
         "title": "Agile Project Management with Kanban",
         "ISBN10": "0735698953",
         "author": "Eric Brechner",
         "pages": 160,
         "publicationDate": "20150326",
         "category": "Software Engineering"
      },
      {
         "title": "The Merriam-Webster Dictionary",
         "ISBN10": "087779930X",
         "author": "Merriam-Webster",
         "pages": 939,
         "publicationDate": "20040701",
         "category": "English"
      }
   ]
}
```

It is important to remember that a single document should only contain the information for that one unique record. This means that you would not want to have two customers in a single document. That would get confusing.

As previously mentioned, documents do not need to be completely uniform. This means that one document can contain properties that another document might not ever have, and both can exist in the same collection. For example, if you had a collection that contained products

## Chapter 1: Fundamentals

for the fictitious bookstore, it would obviously have books in it. It might also have magazines, maps, and puzzles in it as well. The details of these products would need to be somewhat different.

Documents in a collection might each have some common properties such as price, title, description, and weight. For each product type, there would be some unique properties. A puzzle, for example, might have a property that states what the recommended age is for that puzzle.

The following example shows a collection of documents that have both common and unique properties. Note how the document for a book differs slightly from the document for a puzzle, yet both can exist in the same collection.

```
// Bookstore products
{
   "type": "BOOK",
   "title": "Agile Project Management with Kanban",
   "ISBN10": "0735698953",
   "author": "Eric Brechner",
   "pages": 160,
   "publicationDate": "20150326",
   "category": "Software Engineering"
},
{
   "type": "BOOK",
   "title": "The Merriam-Webster Dictionary",
   "ISBN10": "087779930X",
   "author": "Merriam-Webster",
   "pages": 939,
   "publicationDate": "20040701"
   "category": "English"
},
{
   "type": "PUZZLE",
   "title": "Balloons in sky",
   "company": "ZipZap Toys",
   "age": "3-5 years old"
}
```

Just because documents can contain heterogeneous content does not mean that you always want to have collections set up that way. You may decide to have all your collections contain uniformly structured documents. I will later describe circumstances that will help you make these type of design decisions.

*Note*: *I would advise you to stay away from multidimensional properties (an array of arrays). Otherwise, you may find yourself looking through your data and not remembering how you had set it up.*

## A common property of all documents

Each document has an `_id` property. When a document is created, it will always have a unique id associated with it so that it can be identified. This serves as the primary key. You can set the value yourself. If you don't, MongoDB will set a value for you of type ObjectId. If you set it yourself, it can be of any type other than an array. It must, however, be unique across all documents in that collection.

If you ever need to directly access a document, you can access a document using the `_id` as the quickest way to query for it.

## Referencing external data

So far, you have seen that MongoDB allows for storage of JSON documents. As you know, the basic data types available with JSON can be used to represent lots of information you wish to store.

As was mentioned, the one thing that JSON datatypes are not suited for, is the representation of large binary data. For example, you would not really be able to store large media content such as photos, music, or video in a document.

To overcome this limitation, you can create a property that is a reference to where the actual media content is externally stored. You would use a Universal Resource Indicator (URI) to indicate its location, and create a parallel property that describes the type of data it consists of: text, image, binary data, etc. That way, your code can interpret it correctly.

The externally referenced data can be any data you would like. It is up to you to set the type and then treat it as such when you retrieve it. Be sure to handle any needed deletion of your external data if it is supposed to be cleaned up when documents referencing it are deleted.

*Note: This chapter stated that MongoDB, as a document-based database, is not required to enforce any schema. MongoDB does have a capability where you can specify what properties should exist and what restrictions should be on them for documents. I do not utilize this capability in this book. I do later mention that you can use a Node.js module such as Mongoose to schematize your data and/or use joi to validate your schema. The validation MongoDB performs is also not as strict a definition as you have with a relational database. For example, relationship key specifications are available with a relational database with the ability to also enforce what is referred to as referential integrity.*

# Chapter 2: Data Modeling

A data model defines the structure of records that are to be stored in a database. This includes information on how the different types of records relate to each other.

Just because a document-based database is not a relational database, does not mean that it doesn't have structure or relationships between document types. Specifying a data model in MongoDB consists of specifying what the JSON documents are, which collections they exist in, and how they relate to each other. I will instruct you on how to use validation code to make sure any data going in conforms to expectations that conform to your data model.

Sketching out your data model will be an important step in the creation of your data layer. It is helpful to design your document structures in advance so that you can look at all aspects of your data. You can match your data storage needs against the characteristics that a MongoDB database offers and do your data modeling accordingly.

To visualize your data model, you can use whatever diagram format or tool you like. You can even scribble it out on a piece of paper, although you might want ot keep it in electronic form for easier editing. Ultimately, with MongoDB, the JSON format is what you would need to specify. Let's now cover some of the big design decisions that need to be made with a document-based DBMS.

# 2.1 Referencing or Embedding Data

Records in a relational database each contain keys to identify them and to act as references to each other. A relational DBMS has mechanisms to specify and enforce the integrity of these references to some extent.

You may have heard the term "normalization" used in the context of relational database designs. Normalization requires the separating out of data into different record types and relating them to each other. For example, you might have two types of records in your database, one that represents people and another that represents the pets that are owned by the people. There would be a key to relate an owner to one or more of their pets. The following illustration shows this normalized structure:

PART I: The Data Layer (MongoDB)

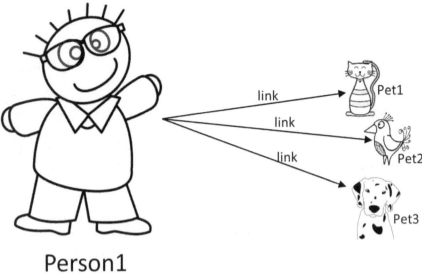

*Figure 8 - Normalized structure*

In a relational DBMS, you set up the keys to relate the record types to one another. The relational DBMS provides functionality to be able to query and join the records together. This means you can query the database to return a person and all the pets they possess. This is done through a single query operation.

Let's now bring this conversation into the MongoDB world. Following a normalized model with MongoDB, you would create separate documents for each person and each pet all in the same collection. Keep in mind that there is no concept of a schema for a collection required, and no concept of relational keys or cross-document join queries. You would need to design your own properties on each document in order to relate pets to their owners. With people and their pets separated out across MongoDB documents, you have achieved a normalized database model.

A normalized model, however, is not necessarily the best way of storing data in a document-based database. You really want to think more about how to denormalize your data. Denormalized data means that everything is all bundled together.

To denormalize, you bundle pets together with a person. You do this by simply placing pets as an array property that is embedded in each person document. Imagine stuffing all the pets in the pocket of the person. This way, people and their pets are always found together. In this way, there are no relational links required. You also do not need a join operation if you denormalize your data. In fact, the classic relational join operation is not even supported in MongoDB. The following visualization shows how the person and pets travel together:

# Chapter 2: Data Modeling

*Figure 9 - Denormalized structure*

As was mentioned, document-based databases like MongoDB do not support joins of records such as cross-table joins done with a relational database. This is because cross-document joins do not make sense in document-based databases. This means that you need to get comfortable with keeping your database designs denormalized. Denormalization is more efficient and works well in a document-based database. What you do, is make use of the array and embedded object properties in your JSON.

In the previous chapter, I showed you a JSON document representing a customer of an online bookstore company. What was shown was actually a denormalized data pattern. Each customer document contained information about the customer, but it also contained information about each book they had purchased. You will end up with larger and more complex documents when you denormalize your data.

Even if you end up with large, complex documents in MongoDB, the querying capability supports retrieval of just the portions of the documents that you need. For example, if you kept pets embedded in a person document, you can do a query to just return the pets of a given owner. You do not need to return the complete person document if you do not want to. Perhaps you just want to know the names of all the pets and do not want to retrieve anything else about the pet or the person.

In summary, it can be stated that database normalization has you separate out your data into different distinct record types and has you set up keys as reference links between them. On the other hand, denormalization, with document-based databases, has you keep as much data bundled together as possible by embedding data that is related.

## 2.2 When to use Referencing

It is obviously not going to work well if you always embed all your data and end up with huge, complicated documents. You need to make decisions as to what properties will be embedded in a document and what would make more sense to pull out into separate document types to reference.

You can keep all of your different document types in one collection. There are, however, reasons why different document types should be kept in separate collections or even separate databases. I will cover more on that topic later.

### Performance implications

Your data model design will have an impact on the performance of all database operations. For example, the performance of reads and writes could improve if you spread data out into separate documents. Doing this, results in decreased data transfer amounts and more efficient document updates as well.

The downside of referencing data from one document to another is that it will be slower if you need to combine data and present it together at any point in time. Imagine the code required to piece together a person with their pets if pet documents were kept separate.

Let's go back to the online bookstore example. Imagine that you have two document types: customers and books. With these documents, you keep track of all the book purchases of individual customers. If you keep only the book IDs in the customer document, certain queries would run slower, such as listing a customer with the book titles they ordered. For each customer, you have to look through the `booksPurchased` array property and then do individual fetches of data for each book id listed. Here is how this normalized model looks:

```
// Customer documents
{
  "_id": "77",
  "type": "CUSTOMER_TYPE",
  "name": "Joe Schmoe",
  "age": 27,
  "email": "js@gmail.com",
  "address": {
    "street": "21 Main Street",
    "city": "Emerald City",
    "state": "KS",
    "postalCode": "10021-3100"
  },
  "booksPurchased": ["0735698953", "087779930X"]
}
```

## Chapter 2: Data Modeling

```
// Book Documents
{
   "_id": "7865",
   "type": "BOOK_TYPE",
   "ISBN10": "0735698953",
   "title": "Agile Project Management with Kanban",
   "author": "Eric Brechner",
   "pages": 160,
   "bookReviews": [
      {
         "reviewer": "Joe Schmoe",
         "comments": "Wish I had this years ago!",
         "rating": 4
      },
      {
         "reviewer": "Jane Doe",
         "comments": "Forever a classic.",
         "rating": 5
      }
   ]
}

{
   "_id": "7866",
   "type": "BOOK_TYPE",
   "ISBN10": "087779930X",
   "title": "The Merriam-Webster Dictionary",
   "author": "Merriam-Webster",
   "pages": 939,
   "bookReviews": []
}
```

Imagine that you want to query and find all books that are over 500 pages in length and that were purchased by a particular customer, Joe Schmoe. To do this, the query must first search the collection to find the Joe Schmoe document and then, for every book ID listed in the Joe Schmoe document, look up that book document and see if the `pages` property value is over 500, and then finally return those documents. This involves more than one document.

In relational databases, this can be done in a single query. In MongoDB, it must be done in separate queries and involves code to piece everything together if needed.

This is not necessarily a bad thing and is just part of what you need to do in a document-based database when you separate out data into related documents. But then again, remember that denormalization relieves you from doing these join operations across documents. You will later see that there is a compromise that can be made between the two options.

# When to reference

Let's cover the basic scenarios that would cause you to split data across documents with a normalized design that uses references. The following are some of the common reasons to do this:

- **The data is rarely used in queries:**
  If you find that you rarely reference certain properties in a document, then you might see this as a sign that those properties belong in a separate document, or even in a separate collection that can be referenced as needed. For example, the address property in the customer document could be taken out and placed into a separate document if it is decided that it is rarely needed. Just remember that you are gaining the faster interaction with the primary data at the expense of slower data retrieval processing to join it together later.

- **The data is common across documents:**
  The book detail is certainly data that would be duplicated across customers. If you had thousands of customers that each had a Merriam-Webster Dictionary, it would not be necessary to keep duplicating all of the data for that book. With duplicated data, one big problem is if one of the properties of a book needs to be altered, it would require altering it across all the duplicates instead of in one central document. In this case, it would be wise to keep the properties in a separate book document if they are shared and updated frequently and make a reference to them.

- **The data has mutual or cross-reference relationship:**
  The information in the `bookReviews` property originates from the customers that submit the reviews, but each review is also specific to a single book. The question arises — should the book reviews exist with the book being reviewed, or with the customer giving them? You might not want to update the individual book document and keep adding to it every time a new customer adds a review. Nor do you want to fetch all of the book reviews every time you fetch the document for a book. This is where you can make the case for the book reviews for each book to be in their own separate document with an array property or have each in individual documents and then referenced by both the customer and the book documents.

- **The data will grow very large:**
  An array property might grow to have a large number of entries. An array property of a document cannot grow unbounded. Remember that there is a 16-megabyte limit on the size of an individual document in MongoDB. You would have to set up several related documents that each had an array that comprised sections of the total set.

Chapter 2: Data Modeling

# 2.3 Reference Relationship Patterns

At one point I showed you a customer document that had a list that contained the ids of the books a person had purchased. This design keeps the documents for the actual book details separated out. This is one of several patterns you can find useful for referencing data.

I will stress again though, that this is totally under your control to implement. MongoDB does not provide any built-in recognition of relationships. It is important to realize that MongoDB will not verify the referential integrity of the data as a relational DBMS might do. It is up to you to manage that. If a given book is deleted from a collection, you also need to delete all the references to it. For example, if a customer document had a `booksPurchased` array property with an entry in it with an id of 5777, there is nothing in MongoDB that is verifying that a book document actually exists with an id of 5777.

Here are some common patterns that can be used for referencing data across documents in MongoDB:

- **One-to-One:** This is where you would separate out a piece of infrequently accessed data. For example, you could pull out the address information of a customer and put it into its own document. This diagram illustrates a one-to-one pattern:

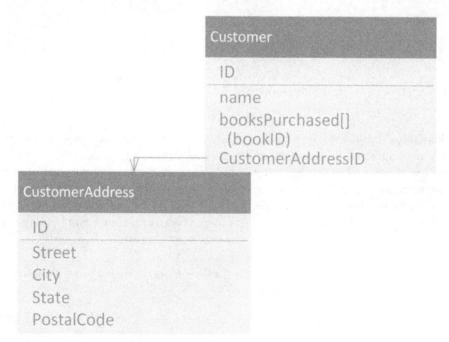

*Figure 10 - One-to-one relationship pattern*

35

PART I: The Data Layer (MongoDB)

- **Many-to-One:** This is where one document might be referenced by many other documents. This is the example you have already seen where a single book can be referenced by multiple customers. Each entry in the `booksPurchased` array has an ISBN10 ID to reference the book document with. This diagram illustrates a many-to-one pattern:

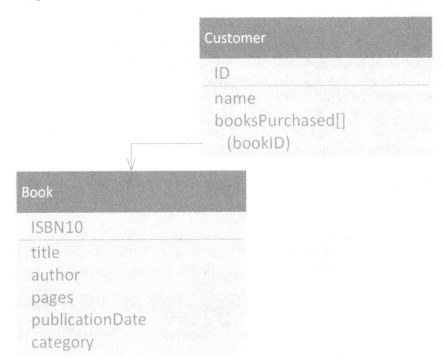

*Figure 11 - Many-to-one relationship pattern*

- **Many-to-Many Association:** This is where you have two separate document types where neither references the other, but instead there is a third document type that can tie the two together through an association. This third document can also contain information relevant to that association. For example, you might have a document for purchases that records the details of the transaction of a book purchase. Thus, you have many purchase documents that reference many customer and book documents. This diagram illustrates a many-to-many association pattern:

Chapter 2: Data Modeling

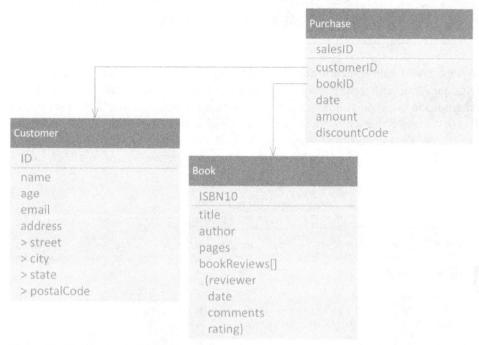

*Figure 12 - Many-to-many association pattern*

Many-to-many relationships could have documents referencing each other in a circular manner. Try not to get that complicated, as it presents many difficult data integrity issues and complicates the service layer implementation.

*Note: It is useful to create a visual diagram of your document types as you go through your data model design process. You are then able to more clearly see the structure and relationships. Several diagram standards have been created over the years to visually represent objects in object-oriented languages and for representing records in databases. All you really need is a simplified visual representation of JSON objects in your data model. You can draw a rectangle shape for each document type and list all the properties inside as I have shown in the previous illustrations. I have chosen to have the ID property exist above a line. Those properties below the line are the rest of the document properties. I have chosen to use a greater than sign to show sub-object properties. An array of objects is shown in parenthesis and it is understood there would be zero or more of these in an actual instance of the document. You might also want to list the data type of each property out to the right of the name.*

## 2.4 A Hybrid Approach

You are now ready to learn about a hybrid approach that can give you the best of both techniques for both referencing as well as embedding data. Why not combine the two techniques in a sort of compromise? For example, if you find that you often need to list the titles of the books that a customer owns, you can duplicate a portion of that information across documents. This means you store the book id and the title, even though it is duplicated information. The following is an example of what the documents would look like:

```
// Customer Document
{
   "_id": "77",
   "type": "CUSTOMER_TYPE",
   "name": "Joe Schmoe",
   ...
   "booksPurchased": [
      {
         "ISBN10": "0735698953",
         "title": "Agile Project Management with Kanban"
      },
      {
         "ISBN10": "087779930X",
         "title": "The Merriam-Webster Dictionary"
      }
   ],
   ...
},

// Book Document
{
   "_id": "77",
   "type": "BOOK_TYPE",
   "ISBN10": "0735698953",
   "title": "Agile Project Management with Kanban",
   "author": "Eric Brechner",
   "pages": 160,
   "publicationDate": "20150326",
   "category": "Software Engineering"
},
```

You can see that the compromise was to keep the title duplicated across document types, as the title is frequently needed. The rest of the properties are kept separated out. Be aware that you would want to synchronize changes to any duplicated properties that existed. For example, if you need to make a change to the `title` property in the book document for a given book, you would want to have some background process that would query all occurrences of that book in the customer documents and update the title in those.

## 2.5 Differentiating Document Types

You might have several different document types existing in a single collection. Storing multiple document types in one single collection allows you to use a single API connection in your code. API connections are tied to a database and for a single collection.

With all document types in the same collection, you would need a way of differentiating them from each other so that queries can find each specific type. One way to solve this is to include a `type` property in each document.

Let me use the previous example of documents for persons and for pets. Person documents would be declared with: `"type": "PERSON_TYPE"` and pet documents would be declared with `"type": "PET_TYPE"`. For example:

```
// Person and Pet types in same Collection
{
   "type": "PERSON_TYPE",
   "_id": 1,
   "name": "Ian",
   "Pet club membership": true,
   "pets": [12, 24],
},
{
   "type": "PET_TYPE",
   "_id": 12,
   "name": "Kirby",
   "breed": "Cavalier dog",
},
{
   "type": "PET_TYPE",
   "_id": 24,
   "name": "Kaitlyn",
   "breed": "Siamese cat",
}
```

This allows a query to narrow down results to just returning the document type you want and then you can add in whatever further criteria you are looking for.

## 2.6 Running Out of Space in a Database

There are ways to deal with cases where the number or size of documents becomes a problem. Think about what would happen if you keep adding documents to a collection that existed on a single SSD (Solid State Disk)? Obviously, at some point, you are going to reach the storage space limit that is set for a single database you are paying for. This does not necessarily mean that you need to separate out and reference data across document types in separate collections. You can still keep data embedded and grow the number of documents indefinitely.

The method for achieving storage capacity scaling is through what is called sharding. This means your collection is more of a logical concept and is actually spread across multiple SSDs. The unit of scaling for MongoDB is called a replica set. A given document, however, must only be found in one replica set (shard) of the sharded cluster. The nice thing is that MongoDB hides that from you, and your query or update does not even realize what is going on.

For customer documents, you could have things set up to distribute customer documents out across different replica set storage. In a later chapter, I will discuss this type of data partitioning. Don't worry if you do not fully understand this concept just yet. You might never need to implement sharding anyway, only if you run into really large amounts of data and need to maintain fast read and write times.

## 2.7 Access Control

There are several ways to interact with MongoDB documents. One is through the Atlas management web portal. Another is through the Compass application, and another is through API access, such as in Node.js JavaScript code. Of course, there is also the Mongo shell, but that is not something that will be needed for the purposes of what this book is teaching. You will learn just the bare minimum to use the mongo shell, such as for importing documents in bulk. If you want to learn more about the Mongo shell, please refer to MongoDB's documentation.

From the Atlas management portal, a user account can be added to provide authentication for that user. You can give a user read-only privileges, if you need that restriction in place. Account administration through the portal is not central to the topic of this book, so if you need more information, please refer to the Atlas management portal documentation.

As mentioned, the primary type of access discussed in this book is done programmatically through a MongoDB API, such as one that is provided for Node.js developers. The details of this will be explained in the Node.js section later in this book.

You will later see how the NewsWatcher sample application has a UI that goes through the middle-tier service layer where interaction with the database takes place. The middle tier will then be able to authenticate on behalf of the user and access the database on their behalf and can restrict access to just documents that a particular user is allowed to see.

# Chapter 3: Querying for Documents

You may have heard of SQL (Structured Query Language) as the language used to query for records in relational databases. Using SQL, you submit queries to your DBMS and receive back the resulting records that match. MongoDB supports querying, but it does not use SQL. Instead, it has its own query syntax.

The lowest level interface to MongoDB has a means of interaction to accomplish operations such as queries. Everything at the lowest level is done through a TCP/IP connection that has a well-defined wire protocol for the operations that it supports. There are around nine operations you can make with this protocol. Don't worry about understanding this, all you need to know is that other people have done the work to abstract away the complexities of using this protocol by creating specific language drivers you can use. This book will be concerned with using the Node.js JavaScript driver. The following diagram shows the overall access layers:

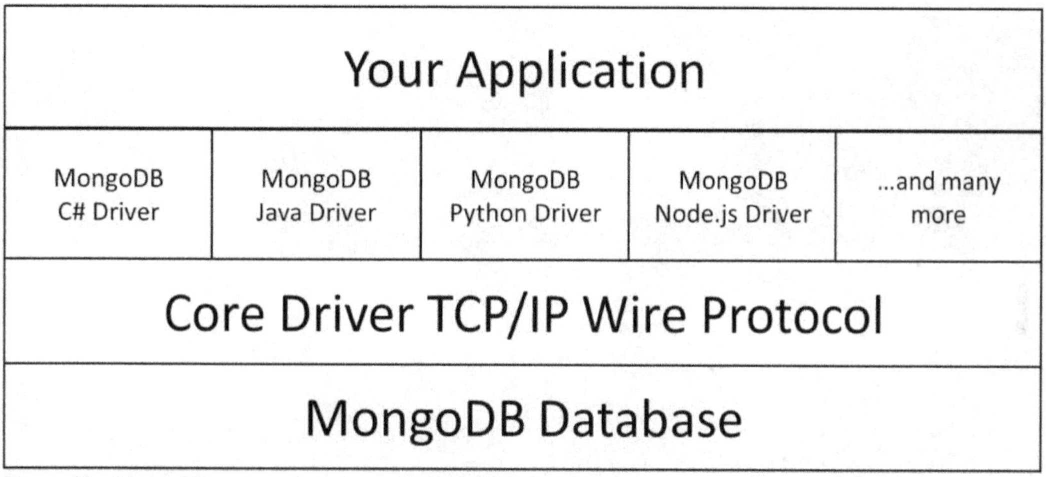

| Your Application | | | | |
|---|---|---|---|---|
| MongoDB C# Driver | MongoDB Java Driver | MongoDB Python Driver | MongoDB Node.js Driver | ...and many more |
| Core Driver TCP/IP Wire Protocol | | | | |
| MongoDB Database | | | | |

*Figure 13 - MongoDB access abstraction through APIs*

In the service layer part of this book, I will show you how to use the MongoDB Node.js driver for operations on the data such as create, read, update, and delete (CRUD). This current chapter will only cover the specifics of the syntax for query operations in general.

Regardless of which of the CRUD operations you perform, you need to specify your query criteria as part of that request. You can explore this topic now, because you don't need to write any code to try out your queries. You can use the Compass application to try them out and experiment with the syntax as you like.

# Example documents

Carefully review the example JSON documents shown below. They will be used with examples showing how to construct your queries. You can imagine these documents being used by an online bookstore. There would obviously be a lot more data available than what is in this example:

```
// Customer Documents
{
   "_id": "77",
   "type": "CUSTOMER_TYPE",
   "name": "Joe Schmoe",
   "age": 27,
   "email": "js@gmail.com",
   "address": {
      "street": "21 Main Street",
      "city": "Emerald City",
      "state": "KS",
      "postalCode": "10021-3100"
   },
   "rewardsPoints": 99,
   "booksPurchased": [
      {
         "id": "1098",
         "title": "Agile Project Management with Kanban"
      },
      {
         "id": "1099",
         "title": "The Merriam-Webster Dictionary"
      }
   ]
},{
   "_id": "78",
   "type": "CUSTOMER_TYPE",
   "name": "Jane Doe",
   "age": 37,
   "email": "jd@gmail.com",
   "address": {
      "street": "100 S Bridger Blvd",
      "city": "Paradise",
      "state": "UT",
      "postalCode": "84328"
   },
   "rewardsPoints": 0
}

// Book Documents
```

## Chapter 3: Querying for Documents

```
{
   "_id": "1098",
   "type": "BOOK_TYPE",
   "title": "Agile Project Management with Kanban",
   "ISBN10": "0735698953",
   "author": "Eric Brechner",
   "pages": 160,
   "format": "Paperback",
   "price": 27.66,
   "publicationDate": "20150326",
   "category": "Software Engineering",
   "bookReviews": [
      {
         "reviewer": "Joe Schmoe",
         "date": "20140321",
         "comments": "Wish I had this years ago!",
         "rating": 4
      },{
         "reviewer": "Jane Doe",
         "date": "20150923",
         "comments": "Forever a classic.",
         "rating": 5
      }]
},{
   "_id": "1099",
   "type": "BOOK_TYPE",
   "title": "The Merriam-Webster Dictionary",
   "ISBN10": "087779930X",
   "author": "Merriam-Webster",
   "pages": 939,
   "format": "Paperback",
   "price": 11.26,
   "publicationDate": "20040701",
   "category": "English",
   "bookReviews": [
      {
         "reviewer": "Joe Schmoe",
         "date": "20100101",
         "comments": "A terrific volume to keep handy.",
         "rating": 4
      },{
         "reviewer": "Jane Doe",
         "date": "20120817",
         "comments": "Wish it came in an audio book format.",
         "rating": 2
      }]
}
```

I will now introduce you to the basics of the query syntax. I won't cover each and every aspect of it. It is fairly robust and much of it is beyond the scope of this book and something you might not ever need to use. For example, I will only briefly mention how to use the MongoDB

## PART I: The Data Layer (MongoDB)

aggregation features and its syntax. For more information on that topic and other supported syntax intricacies, refer to MongoDB's documentation.

## Syntax overview

In a later chapter, you will be querying a collection using a MongoDB NPM module to write code in JavaScript that runs in Node.js. That module provides functions such as `find()` or `findOneAndDelete()`. The first parameter of those functions is the query criteria that specifies the matching to take place across all of the documents in a collection.

If you are familiar with SQL, this is similar to what a WHERE clause does. Here is an example query using the `find()` function with a greater than operator in the query criteria. This query below will return all of the documents in the collection whose age property contains a value greater than 35. This is showing you how you would call it in code, and this also happens to be the format you use in the Mongo shell.

```
db.collection.find({age: {$gt: 35}});
```

For the function call above, there is the possibility that the query criteria won't match anything. This is ok, and no documents will be returned. On the other hand, if there are a lot of documents returned, then you need to use the Node.js driver capability to fetch results in batches. You will see how to actually access the results of the `find()` function in code.

Besides the query criteria I just showed you, there are also criteria you can provide for what is called the projection criteria. The projection criteria determine the properties that will be returned from each document. Here is another example using the same `find()` function, but this time with an optional second parameter that specifies the projection:

```
db.collection.find({age: {$gt: 35}}, {name: 1, age: 1});
```

This query will find all documents in the collection whose age property contains a value greater than 35. With the projection specified, only the name, age and _id properties will be returned. _id is always returned, unless you specify otherwise.

As mentioned, the second parameter in this query is the projection criteria. This is where you list the properties you want to be returned. The number following the colon determines if the property is included, or if it is excluded. I will explain more about this soon.

I will keep using the example of the bookstore customer document, but for now, just pretend they only have four properties each (_id, name, age, and email). The following diagram of the previous query shows the two criteria in the function call and how they determine the output:

## Chapter 3: Querying for Documents

**Customer Documents**

```
"_id": "77",
"name": "Joe Schmoe",
"age": 27,
"email": "js@gmail.com"

"_id": "78",
"name": "Jane Doe",
"age": 37,
"email": "jd@gmail.com"
```

Query criteria `{age: {$gt: 35}}` →

```
"_id": "78",
"name": "Jane Doe",
"age": 37,
"email": "jd@gmail.com"
```

Projection criteria `{name: 1, age: 1}` →

```
"_id": "78",
"name": "Jane Doe",
"age": 37
```

*Figure 14 - Criteria flow*

You can see that the query criteria determine what documents pass through to the result set. The projection criteria select what properties you want for each document in the result set.

Now I can go into the details of both the query and the projection criteria operations. You can connect to your MongoDB hosted cluster through the Compass application, add a database, add a collection and some documents, and then try out some queries on your own.

*Note*: *You cannot just create a query and assume that it will end up being efficient. The execution time of a query can vary greatly. To address performance issues, you either must create indexes that can speed up your queries or think about a more efficient way of modeling your data. The topic of index creation is covered later.*

# 3.1 Query Criteria

The query criteria are actually optional on a function such as the `find()` function. It is, however, something you will almost always be using. If you call `find()` without any query criteria parameter, every single document in the collection will be returned.

The query criteria are tests that are applied to the collection to see what documents are to be included as part of the result set. If you really want to, you can narrow down the result to return a single document. For example, you can query by the _id property with an equality test. The _id property is unique, so each document can be uniquely identified with it. Here is an example of code to query by _id. The result it returns is also shown:

```
// Query
db.collection.find({_id: {$eq: "77"}}, {address.state: 1});

// Results
{
   "_id": "77",
   "address": {
      "state": "KS"
   }
}
```

PART I: The Data Layer (MongoDB)

You don't need to write any code to try this out, but can use the Compass application to try out queries. You can go to chapter 6 to learn how to create your PaaS hosted MongoDB cluster, database, and collection. You will also find information on installing the mongo shell. You can use that knowledge to import documents to use as you experiment with queries. See the Example Documents section, a few pages previous, for the documents to create.

Once you have your test database, test collection, with documents imported, you can launch Compass and go to your collection and in the Documents tab, enter some test queries. Here is what the Compass UI looks like if you are trying out a query:

*Figure 15 - Compass application querying capability*

Each query criteria can utilize one or more operators. The example above uses the $eq operator. The real power of the query is in the use of the criteria operators. I'll now go over what those are and show you some examples.

## Criteria operators

You have seen operators such as $gt and $eq used in the examples in this chapter. Those operators stand for greater than and equal to. There are many more operators that you can use in your query criteria to filter documents. The following operators are currently supported:

- **Comparison**
    - $eq
    - $gt
    - $gte
    - $in
    - $lt
    - $lte
    - $ne
    - $nin
- **Array**
    - $all
    - $elemMatch
    - $size
- **Bitwise**
    - $bitsAllClear
    - $bitsAllSet
    - $bitsAnyClear
    - $bitsAnySet
- **Element**
    - $exists
    - $type
- **Evaluation**
    - $expr
    - $jsonSchema
    - $mod
    - $regex
    - $text
    - $where
- **Geospatial**
    - $geoIntersects
    - $geoWithin
    - $near
    - $nearSphere
- **Logical**
    - $and
    - $or
    - $nor
    - $not

Each of the operators uses its own unique syntax. The `$eq` operator uses the following syntax:

`{<name>: {$eq: <value>}}`

The `name` is what you want to test against. It can be a top-level property, or it can be a property within the hierarchy of the JSON. The `value` is a string, number or other value that matches the data type of the property. Here is an example that tests a second-level property; referred to as an embedded document field in MongoDB documentation:

```
{"address.state": {"$eq": "UT"}}
```

You can even specify a property of an element found in an array. If you look at the example document, you see that `booksPurchased` is an array and `id` is a property of each of the elements of that array. In the example below, the returned document is the complete document, as the query criteria are only used to find a document match and not specify properties to return.

```
{"booksPurchased.id": {"$eq": "1098"}}
```

Here are a few operators from some of the categories to give you an idea of how they work. For more detailed information on each of the operators, see MongoDB's documentation.

## Comparison

With the comparison operators, you need to first select the property name you are interested in testing against. This is then followed by the comparison operator and finally the value you want for that comparison test. The only exception to this is with the `$in` and `$nin` operators, which use an array and not a single value.

I will use the same example I have shown you previously, which is doing a query for a single document by its _id value:

```
{"_id": {"$eq": "77"}}
```

To just test the equality of a property, you can shorten the syntax to the following:

```
{"_id": "77"}
```

You can use more than one comparison operator at a time, such as you would need to do to test ranges of values. All operator tests need to pass their test in order for a given document to be included in the results set.

Here is an example that queries for books that have between 100 and 200 pages:

```
// Query
{
   "pages": {
      "$gt": 100,
      "$lt": 200
   }
}
```

# Chapter 3: Querying for Documents

**Logical**

The logical operators let you string together several tests in a row to perform the desired logical testing. The syntax for the logical operator $and is:

{$and: [ { <expression1> }, { <expression2> } , ... , { <expressionN> } ]}

The logical operator syntax starts with a boolean operator such as $and. It then contains an array of expressions that can be made up of individual comparison operators that we have seen previously.

The following query looks for books that are less than 200 pages in length and which are also in the Software Engineering category.

```
// Query.
{"$and": [{"pages": {"$lt": 200}},
          {"category": {"$eq": "Software Engineering"}}]}

// Results. Assuming you also have a projection criteria of {"_id": 1}
{
    "_id": "1098"
}
```

If you are only carrying out this one level of boolean operation, then you don't really need the $and operator. Instead, you can just list the conditions one after another. Here is the same example query without the $and operator:

```
// Query.
{"pages": {"$lt": 200}, "category": {"$eq": "Software Engineering"}}

// Results. Assuming you also have a projection criteria of {"_id": 1}
{
    _id": "1098"
}
```

Here is a query that uses both the $and and the $or operators. I'll use the complete expanded text formatting of the query as it is easier to read. This example queries for books that have less than 200 pages and are either in the category of Software Engineering or Science Fiction.

```
// Query.
{
  "$and": [
    {
      "pages": {
         "$lt": 200
            }
    },
    {
```

49

## PART I: The Data Layer (MongoDB)

```
      "$or": [
      {
        "category": {
           "$eq": "Software Engineering"
                    }
      },
      {
        "category": {
           "$eq": "Science Fiction"
                    }
      }
              ]
         }
      ]
}

// Results. Assuming you also have a projection criteria of {"_id": 1}
{
   "_id": "1098"
}
```

If you find your query has a lot of $or operations to match on many different values for the same property, then you can use the $in operator. The value to match can even be a regular expression. The following example shows how easy it is to use this to list all the possible matches:

```
// Query using IN
{
   "category": {
     "$in": [
        "Software Engineering",
        "Science Fiction"
     ]
   }
}
```

### Element

The element operators $exists and $type are for selections based on whether a property exists and if it is of a certain datatype. The following example shows the syntax for $exists:

```
{ name: { $exists: <boolean> } }
```

A typical use for $exists would be to use this operator inside another operator. In this example, you want to make sure the document has the publicationDate property. It may be that a book has a price set, but has not been published yet, so the publicationDate property is not there yet.

```
{"$and": [{"price": {"$lt": 30}}, {"publicationDate": {"$exists": true}}]}
```

## Evaluation

If you struggle to get exactly what you want in your query, you might find the evaluation operators are just what you need. Here is an example that shows the use of a regular expression with the option for a case-insensitive test. This example will find all books that have a title that starts with the word "agile" no matter the letter casing:

```
{
   "title": {
      "$regex": "^agile",
      "$options": "i"
   }
}
```

You might have a document with large amounts of text that you want to search to see if specific words or phrases exist. You can use the `$text` operator in this case. To use this, you need to first create an index of type text on the properties you want to use it on. For example, you could create the index on the `title` property and then search for books that contain certain words in their title.

```
{"$text": {"$search": "MongoDB"}}
```

For those rare occasions where you just cannot get what you want with any of the available operators, you can resort to writing JavaScript using the `$where` operator. Here is a test that checks to see if a person has purchased more than a single book. This requires JavaScript because the length property of the `booksPurchased` array is only accessible through the API returned object.

```
{"$where": "this.booksPurchased.length>1"}
```

You cannot presently try this query out from the Compass app if your hosting is set up to use the Atlas free tier. The same is true for your code.

# Object and array properties

So far, the examples shown have been testing properties that are single values, such as a string or a numeric data type. But what if you have a document that has a property that is an array of strings? What if you have an object property? Even better, what if you have a property that is an array of objects?

The `address` object property in the bookstore customer example document ends up as an embedded document in MongoDB BSON. You can do a search for an exact match on an embedded document and specify individual names to match as shown in a previous example.

For arrays, you can do an exact match on the full contents of the array, or just on specific values existing somewhere within the array. If the array holds objects, you can search for the element entry and sub-property off of that. Here is a previous example query that was doing this:

```
{"booksPurchased.id": {"$eq": "1098"}}
```

Just as you can test multiple single property values, you can also do that for properties that are arrays objects. Let's say you wanted to search for books with book reviews by Joe where he gave a four-star rating. Here an example of how that would look:

```
{
   "bookReviews.reviewer": "Joe Schmoe",
   "bookReviews.rating": 4
}
```

For a simple array of strings, you could match for that exact array. To search for documents where one string entry in an array exists, you could do an equality test. Here is an example document with a property that is an array of strings:

```
{
   "favoriteColors": ["green", "red", "blue"]
}
```

To include that document, here are the query criteria you could use:

```
{"favoriteColors": "green"}
```

Array searches and projection capabilities in MongoDB are very powerful. If you take the time, you can learn how to match on things like an element in a specific index, or do something like return the first numeric element that is larger than some value. You will have to learn that yourself, as it is tricky to explain all the nuances. See the MongoDB documentation. For example, look at the $elemMatch operator documentation.

## Data type mismatch problem

Equality comparisons can end up producing an undefined outcome if the property data type specified does not match up with the test value data type. For example, you cannot test for a number value on a property that is a string.

The test statement syntax of MongoDB does not work the same as it does in the JavaScript language. The following query will not work because of the data type mismatch:

```
// The selection will not work, as the _id property is a string
// in the document, and you are comparing it with a number
{
   "_id": {
      "$eq": 77
   }
}
```

With the following JavaScript code sample, you can see that data type coercion happens. A boolean test between a string and a number actually works in JavaScript.

```
// JavaScript uses "==" for equality testing.
// Coercion rules apply and both equality tests evaluate to true
var v = "77";
v == "77"; // true result
v == 77;   // true result as coercion happens
```

## 3.2 Projection criteria

Just because you have your query criteria returning the proper result set of documents does not mean that you are done. You may also want to set up projection criteria to just return the properties that you really need. You have seen this demonstrated already, but now you can look at this in more detail.

There may be cases where documents with all of their properties are what you actually want returned. This may be the case with a very sparse document, making it reasonable to return the whole document every time. With larger, more complex documents, you can benefit from restricting the properties being returned. Limiting what properties are returned saves on the amount of data transferred.

### Inclusion and exclusion

You have already seen a projection criteria in use, so you really know most of what you need to know already. Just to review, if you don't provide projection criteria, then the complete document is returned. If you do provide projection criteria, then you can specify the inclusion or exclusion of whichever properties you would like. Exclusion means to return all properties except the ones you list. The inclusion and exclusion syntax is as follows:

```
<name>: <1 or true or 0 or false>
```

True means to include and false means to exclude. You cannot mix both inclusion and exclusion in the same projection criteria. The only exception to this is if you are using inclusion criteria, you can also specify one single exclusion if it is to exclude the _id property.

Here are some examples of different projection criteria with a comment added to state whether they are valid or invalid:

```
{"address.state": 1}       // Valid
{"address.state": 0}       // Valid
{"name": 1, "age": 1}      // Valid
{"name": 1, "age": 0}      // Invalid
{"age": 0}                 // Valid
{"name": 1, "_id": 0}      // Valid
```

As I mentioned, this is extremely handy. Let's say you want to create a list of people with their addresses. You could use the following projection criteria:

```
{"name":1,"address":1,"_id":0}
```

The following is the result set returned:

```
{
   "name": "Joe Schmoe",
   "address": {
      "street": "21 Main Street",
      "city": "Emerald City",
      "state": "KS",
      "postalCode": "10021-3100"
   }
}
{
   "name": "Jane Doe",
   "address": {
      "street": "100 S Bridger Blvd",
      "city": "Paradise",
      "state": "UT",
      "postalCode": "84328"
   }
}
```

## Missing properties

Since MongoDB can be schema-less, it is possible that any number of documents in a collection that you are querying might not even contain the given property that you have specified in your selection criteria. For example, it is possible that `booksPurchased` is a missing property in some of your documents, by your own design. This is important to consider when you are constructing your query criteria and projection criteria.

The following example query will return two documents, but the second document will not have the `booksPurchased` property. This is because the second customer has not bought any books yet.

```
// Projection criteria
{ "name":1,"booksPurchased":1}

// Results
{
   "_id": "77",
   "name": "Joe Schmoe",
   "booksPurchased": [
      {
         "id": "1098",
         "title": "Agile Project Management with Kanban"
      },

      {
         "id": "1099",
         "title": "The Merriam-Webster Dictionary"
      }
   ]
}
{
   "_id": "78",
   "name": "Jane Doe"
}
```

If you really want this second document left out completely, if the property does not exist, use a query selector to only get those with a non-null value as shown here:

```
// Query criteria
{
   "booksPurchased": {
      "$ne": null
   }
}
```

Of course, the property could still exist and just be a zero-length array and it would be returned in that case.

## Arrays

There is a special operator named `$slice` that allows you to return just specific portions of array properties. Examine the following document that has a property containing an array of colors:

```
{
   "favoriteColors": ["green", "red", "blue"]
}
```

# PART I: The Data Layer (MongoDB)

Here are a few examples of the use of different operators like `$slice`, `$,` and `$elemMatch` to pull out different elements from the array property shown above:

```
// Return first two elements
{ favoriteColors: { $slice: 2 } }

// Return first element
{ favoriteColors.$: 1 }

// Return first element that matches
{ favoriteColors: { $elemMatch: { $eq:  "red"} } }
```

For more information on these operators, see MongoDB's documentation.

# 3.3 Querying Polymorphic Documents in a Single Collection

In the section on data modeling, I mentioned that you may decide to normalize some of your data. You might like to store data in completely different document types in the same collection. To do this, your query must always include a way to pick out just the documents of a particular type that you want to have returned.

Using the bookstore example, if you had the customer and the book documents in the same collection, you could include a `type` property in each.

The following is an abbreviated example showing this approach:

```
// Customer documents
{
   "_id": "77",
   "type": "CUSTOMER_TYPE",
   "name": "Joe Schmoe",
   ...
}
{
   "_id": "78",
   "type": "CUSTOMER_TYPE",
   "name": "Jane Doe",
   ...
}
```

## Chapter 3: Querying for Documents

```
// Book Documents
{
   "_id": "1098",
   "type": "BOOK_TYPE",
   "title": "Agile Project Management with Kanban",
   ...
}
{
   "_id": "1099",
   "type": "BOOK_TYPE",
   "title": "The Merriam-Webster Dictionary",
   ...
}
```

In every query, you would need to include query criteria that were specific to the type of document you needed. Here is an example of that query criteria:

```
// Query criteria for retrieving all books
{"type": "BOOK_TYPE"}
```

# Chapter 4: Updating Documents

The previous chapter covered the topic of querying or the 'R' for Read operations in the CRUD acronym. This chapter will cover the 'U' for the Update operation. I will give an overview here of how the syntax works.

When updating a document, you can certainly provide the complete document for uploading. As an enhancement, MongoDB allows you to do things like specifying a single property to be updated. There are many update operators you can choose from and they can be combined in a single atomic update submission for a given document.

To use the update capability of MongoDB, you first need to provide the query criteria to identify the document or documents to be updated. That criteria uses the same syntax as already covered for the query criteria. What is new here is that the update criteria is added as another parameter. Here is the update criteria syntax:

```
{
   <operator1>: { <name1>: <value1>, ... },
   <operator2>: { <name2>: <value2>, ... },
   ...
}
```

*Note: Create and Delete operations are being skipped. Creation is just providing the JSON document to create and a deletion operation uses the same query criteria syntax that a read does. All MongoDB CRUD operations will be covered in the service layer discussion.*

## 4.1 Update operators

The following are the currently supported update operations used on single value properties:

- `$currentDate`
- `$inc`
- `$max`
- `$min`
- `$mul`
- `$rename`
- `$set`
- `$setOnInsert`
- `$unset`

Chapter 4: Updating Documents

Here are some examples to illustrate some of these update operators. I will only show examples that update a single property at a time. You can combine multiple operators on different properties in one update submission. These will all be committed at the same time and will result in an atomic operation at the document level.

There are many operators you can use, and I will give examples of a few of them. For every update call, you need as the first parameter the query criteria to identify the document(s). The second parameter specifies the property to update. If the query criteria identify more than one document, the update happens on all of those documents identified.

*Note: The Compass app does not allow the use of the update syntax right now, only query searches work. That is why I am showing the examples with code.*

# $set

The `$set` operator is the standard way to replace the value of a property. The following example sets the `rewardsPoints` property to a new value for the queried person:

```
db.collection.update({_id: "77"}, {$set: {rewardsPoints: 1000}});
```

If the property did not exist in the document, it is created. This operator can be used to do a complete replacement of any value, even doing a replacement of a complete array or an embedded document. It also works to replace a specific property of an embedded object.

`$rename` can be used to give a property a new name. `$unset` will delete a property.

# $inc

The `$inc` operator is used to change the integer value of a property by a specified amount. You can add or subtract from any value. Using the bookstore example, you could add rewards points to a customer. Here is an example of what that would look like:

```
db.collection.update({_id: "77"}, {$inc: {rewardsPoints: 10}});
```

# $min and $max

The `$min` and `$max` operators are used to test a given value and only replace it if the value is less than or greater than the test value, depending on the operator. Here is an example:

```
db.collection.update({_id: "77"}, {$max: {rewardsPoints: 2000}});
```

In this example, if the `rewardsPoints` property had a value of 1000 to start with, it has a value of 2000 after this update.

# 4.2 Array Update Operators

The following operators are for use with array properties to perform updates:
- `$`
- `$[]`
- `$[<identifier>]`
- `$addToSet`
- `$pop`
- `$pull`
- `$pullAll`
- `$push`

## $push

The `$push` operator is used to add another element to an array property. Here is an example that adds a new book review to a book document:

```
db.collection.update({_id: "1098"},
  { $push: { bookReviews: {
    reviewer: "Skylar",
    date: "20150923",
    comments: "It was really profound!",
    rating: 5
  }}});
```

The `$addToSet` operator is similar to `$push` except it checks to see if an identical entry exists already and only adds the new element if it is not already present in the array. You can use the additional operator of `$each` to add multiple elements at once. The `$sort` operator can be combined with the `$push` and `$each` operator to keep the array sorted. Combine the `$position` operator with `$push` to specify the point of insertion.

## $

The `$` operator is used to specify that the update is to happen for only the first element of an array that is found to match the query criteria. The part of the `$set` that uses `bookReviews.$.rating` uses the `.$` to signify that the replacement should take place on just the first element match that is found. This example does a search for any document that has a `bookReviews` element that has a rating of 4 and then updates the first matched element:

```
db.collection.update({_id: "1098", bookReviews.rating: 4},
    { $set: { "bookReviews.$.rating" : 5 } }
)
```

## $pop

The $pop operator is used to remove elements from an array property. This example removes the first book review:

```
db.collection.update({_id: "1098"}, {$pop: {bookReviews: -1}});
```

You can use $pull to remove all entries from an array that match what you specify. You can use $pullAll to specify more than one match for the removal criteria.

# 4.3 Transactions

A single write operation might modify multiple documents if the selection query matched more than one document. The modification of each individual document is atomic, but it is not atomic across all of the modified documents. It is possible to isolate a single write operation across multiple documents using the $isolated operator, but only if there is no sharding involved.

You might also have a need to modify two documents simultaneously, and even modify different properties on each. If you have a requirement to change two documents simultaneously, then you need to work this out on your own. For example, if you wanted to take rewards points from one document property and add them to a different document, this would require what is termed a multi-document atomic transaction.

This is a new capability in MongoDB 4.x. Previously, you would need to build that capability yourself. Search "two-phase commit" as per MongoDB's documentation.

You can imagine how important this would be to get right in an application that manages financial transactions across financial accounts. MongoDB now supports multi-document ACID transactions. You should only turn this on if you really need it. The point is that a document-based database might not need this in most circumstances. In the code, you would use an API call to start a session that will surround several operations that you want to be transacted. Then you would either abort it or commit it. Basically, something like this:

```
s = client.start_session()
s.start_transaction()
…code to do multiple updates, inserts etc. on different documents.
---> on an error exception you would call s.abort_transaction()
---> on success you would call s.commit_transaction()
s.end_session()
```

# Chapter 5: Managing Availability and Performance

In an ideal world, you could store an infinite amount of data, access it from anywhere in near-zero time and never have any data loss or corruption. Reality is that it takes a lot of work to approach these ideals. As you design your data model and subsequently try it out, you need to tune your DBMS for consistency, availability, and performance. You can now consider what mechanisms are at your disposal to approach these ideals.

It really takes a fair amount of time to fine-tune each aspect of the management of your MongoDB database. Many times, you will be faced with tradeoffs that have to be made. This chapter will look at some of the aspects that can be "fine-tuned" for specific access scenarios.

In a PaaS environment, some of this work should be less than has been traditionally required in the past with a DBMS. MongoDB has certainly done a great job of making some things automatically happen that used to require a lot of manual configuration.

## 5.1 Indexing

Imagine you had a problem finding personal belongings such as your car keys, the TV remote, your favorite pair of socks, your wallet or purse, etc. Perhaps when you try to head out the door, you find yourself frantically searching for your car keys every morning.

One approach to solving this would be to keep a whiteboard right next to your front door that had two columns. One column would list the item you cared about and the other column would list the known location of that item. The whiteboard might look like this:

| Car keys | Left side pocket of the jacket hanging in the entryway closet |
| TV remote | In the pile of toys in the room of your two-year-old toddler |
| Purple socks | Laundry room floor under the pile of towels |
| Wallet | Under the couch cushion in the TV room |

Imagine the huge time savings this could provide. I saw one study that stated that, on average, a person spends a whole year of accumulated time looking for lost items during their lifetime.

A database index uses the same concept as the whiteboard look-up table, with the goal of serving database records faster. A database index works by creating a separate lookup list that allows for faster querying. This alleviates the need to search through all documents to find the one(s) you are looking for. For example, let's say you had a lastName property in every

# Chapter 5: Managing Availability and Performance

document of a collection. If you created a query that was looking for a particular person with the last name of "Smith", how would a query find it as quickly as possible? The slowest way to search would be to start looking at all the documents one by one, until a document was found with "Smith" in the lastName property. That type of search has no choice but to search each and every document in an unsorted storage system. In a huge collection, this would be a major performance problem.

Indexing can speed up your search by creating a separate sorted list of last names to search against. Each entry would point to the corresponding complete document. A query for the last name of Smith quickly finds those entries with something like a binary search.

As an example, here is a representation of random documents. There is one row per document that exists in a MongoDB collection. This is only an abstract representation of how it is stored. To search for "Tuttle," you would start a sequential search from document to document until you found a match on the lastName. Unfortunately, in this case, it would be the last one found.

## Customer Documents

| lastName | zipCode | rewards | age |
|---|---|---|---|
| Smith | 27896 | 77 | 32 |
| Williams | 43890 | 3 | 18 |
| Adams | 99054 | 654 | 55 |
| Tuttle | 12345 | 567 | 21 |

*Figure 16 - Customer documents unsorted and with no index*

If you add a second list that contains the `lastName` property. Along with the name, you could have a link to the customer document. This second list can be sorted, and your searching will be much faster. A quick binary search of the index list will find the `lastName` to match and then use the link to get the document.

Implementing a good index can be a critical part of your work to maximize the efficiency of your queries. It is well worth your time to measure and analyze the performance of your database and to fine-tune it with indexes. There are reports in the paid Atlas tier subscriptions that you can bring up that help you look at index performance.

The following example illustrates the concept of linking through an index.

## Index

| lastName | link |
|---|---|
| Adams | 3 |
| Smith | 1 |
| Tuttle | 4 |
| Williams | 2 |

## Customer Documents

| lastName | zipCode | rewards | age |
|---|---|---|---|
| Smith | 27896 | 77 | 32 |
| Williams | 43890 | 3 | 18 |
| Adams | 99054 | 654 | 55 |
| Tuttle | 12345 | 567 | 21 |

*Figure 17 – Last name index for Customer Document querying*

You can use the Compass tool for creating all your indexes. There are ways to do this programmatically, but for areas of the application that only need to be set up once, I always prefer to do this in Compass.

***Note:*** *Index configuration can get rather complex and I can only cover the basic common scenarios pertaining to the sample application. You will have to go to the online documentation to get all the details on what is possible.*

# Single-property index

The simplest way to learn about indexes is to learn how to set up an index on a single property. This section goes over how this works for a single value or object property. The next topic explains the subtleties of what happens if the property is an array data type.

Let's go back to the example of the online bookstore. If you look at the requirements for your querying, you can see that you need to be able to search for customers by name. If you had hundreds of thousands of customers, it is certainly going to improve the performance of this query if you create an index on the name property.

The syntax used to create an index is similar to the syntax used to set up your other criteria. This example shows how to specify an index on `name`:

```
{"name": 1}
```

That is how simple it is. You can also add an index on a property of an embedded document. For example, what if you wanted to look up all customers that resided in a certain postal code? To make this query run faster, you would want to place an index on the sub-property as follows:

```
{"address.postalCode": 1}
```

# Chapter 5: Managing Availability and Performance

You could place an index on the address property as a whole, but then you would have to put in a complete address for the query criteria, including having the properties in the same order for the match to succeed.

*Note*: *The _id property that exists on every document is automatically indexed, so you never need to add an index for that.*

## Array property index

If a document property is an array, you set up an index on it in the same way as for a single value property. There are just a few restrictions. One restriction is that you cannot have more than one array property index at a time. For example, if your documents have multiple array properties, you can only set up an index on one of the array properties.

The index would then be used to match on anything found in the array. The index can only be used for individual element matching and not for matching on the array as a whole. As an example, let's say you had the following documents:

```
{"name": "Kiara", "favoriteColors": ["blue", "yellow", "cyan"]}
{"name": "Tristan", "favoriteColors": ["green", "red", "blue"]}
{"name": "Halli", "favoriteColors": ["juju", "nana", "mango"]}
```

You can set up an index similar to the previous example:

```
{"favoriteColors": 1}
```

Now you can set up a query with query criteria to match a color that might be found in the array. If you search for "red," then the document for Tristan will be included in the results set. What is happening internally is that the index contains all array element entries across all documents. There would be three entries in the index for Kiara, alphabetically sorted by the color name. Each of those three would point back to the one document. Similarly, there will be three entries for Tristan and Halli.

You can also set up an index even if the array contains elements that are objects, as with the following example:

```
{"name": "Skylar", "clothes": [
   {"type": "school dress", "color": "tan", "size": 5},
   {"type": "school pants", "color": "tan", "size": 6},
   {"type": "church shoes", "color": "white", "size": 3}]
{"name": "Korver", "clothes": [
   {"type": "pajamas", "color": "green", "size": 3},
   {"type": "tennis shoes", "color": "brown", "size": 2},
   {"type": "winter coat", "color": "blue", "size": 3}]
```

The following example will set up an index just for one property on the object in the array:

```
{"clothes.color": 1}
```

## Multiple-property index

Many times, you will have queries that specify more than one property to match on. This is used when you want to narrow down your search even further. What if you wanted to run bookstore specials to encourage people to use rewards points.

For example, what if you were looking for all people less than 20 years old that had no rewards points so you could give them some points as a free bonus offer to try them out. The following example creates an index on both `age` and `rewardsPoints` so you can search for customers that way:

```
{"age": 1, "rewardsPoints": 1}
```

This is called a compound index. Be sure to understand how this really works. The underlying index has sorted entries for `age` and then, has sub-sorted entries by `rewardsPoints`. This means that you cannot use this index for a query for just rewards points. Finding all people with over 1000 points will not be able to use the index. You can, however, use this index to query just by age. This means that the order of properties listed for the index is important.

You could create two separate indexes to be able to query by both age and rewards points separately as well as combined in either order. MongoDB will use something called an index intersection for you if it can. For example, you could create the following two indexes:

```
{"age": 1}
{"rewardsPoints": 1}
```

With this configuration, you can still query by the combination of age and rewards points. But you can also query just by age, or just rewards points. You can also reverse the combination and query by rewards points and then age, in that order. The only downside is that each index you set up means more storage overhead.

## Index sort order

Up to this point, I have always been using the numeric value of `1` in all of the index examples. What that is doing is instructing MongoDB to create the index in ascending sort order. You can alternatively specify `-1` and get a descending sort order. For example, you could create the following index:

```
{"age": -1}
```

Chapter 5: Managing Availability and Performance

For single-property indexes, this does not matter. It only matters if you want to return multiple documents of more than one property through a corresponding multi-property index and you want the result returned in a specific sort order. For example, querying for all people with the last name of "Smith" and returning the documents in descending order by age. Otherwise, you can certainly add a command to sort the result that is returned.

With the API usage, there is a `sort()` function that can follow the `find()` function that can process the sort order for you, but it will be done outside the index and be slower.

## Other types of indexes

The index examples used so far would be used for exact value matches and range type queries. Just so you are aware, there are a few other types of indexes that you can investigate and utilize that may suit your needs. Some of these other index types supported by MongoDB are geospatial, hashed and text indexes.

For example, let's say you were keeping a database of restaurants along with their menus, reviews, and location coordinates. With a geospatial index, you could then perform queries to take a person's current location and find all restaurants within a certain radius of them.

If you have large amounts of text in a property, you can use what is called a text index. For example, if you were storing news stories and you wanted to have an index on the story content, you could find a text index very performant.

There is an index called a hash index. It can be used on properties that hold a single value (i.e. string, number) or that have an embedded document, but it cannot be used with array properties. The hashed index is populated with the hashed values. A hash index by itself does not work for range-based queries, so you might want another separate single-property index for that.

For embedded documents, the index gives a hash of the complete embedded document. This then would potentially be a faster lookup as the search is just comparing a hash value to find the document. A hash index might make more sense when combined with sharding. This will be covered in an upcoming chapter.

## Index creation options

One of the options you can use with the creation of an index is to specify that the values for a given property are unique. For example, in the bookstore example, each customer has an `email` property. You could make the index require that all emails must be unique across all documents. Here is an example that creates an index with the `unique` option:

```
{"email": 1}, {unique: true}
```

You should create an index like this before any data is populated. Index creation would fail if multiple documents exist that have the same value for their email property. Once an index is created with the unique option, any document creation will fail unless it has a unique value on the indexed field with this option set.

The `sparse` option can be used to create an index that only has entries for documents that contain that property. In the following example, if a document did not have the `email` property, then it would not be included in the index. Any subsequent query for values to match on using the `email` field would ignore those documents, but that is probably what you wanted. If an index exists for a property, then MongoDB will use it, so that is why the rest of the documents would not be searched that are missing that property.

```
{"email": 1}, {sparse: true}
```

## Index creation options

That about wraps up the topic of indexes. As I stated at the start of this chapter, there are no perfect solutions in the world of databases. This is true with indexing. What you need to do is to completely understand what your query needs are first in order to understand how best to create your indexes.

For example, you might even consider not having any index under certain circumstances. If you had a database with 95% of the operations being writes and 5% were reads, you might not want to create an index. This is because an index will slow down your write operations. There are, however, configuration settings you can still make to speed up writes.

If you have the opposite situation, and read performance needs to be fast, and writes are a small percentage of the load, then definitely create indexes. You can always measure your performance before and after to make sure that your indexing decisions are valid.

*Note: Once your database is up and running and you are running code queries against it, you can get a diagnostic report on an individual database. This will tell you how well your index is performing. This is done by either using the* `explain` *option or the* `explain()` *method, depending on the API call you are making.*

# 5.2 Availability through Replication

A single point of failure is never good, no matter what service you are using. For example, in the days of the telegraph, there might have only been a single telegraph line connecting two cities. That configuration creates a service that has a single point of failure. Cut the single line and communication is severed. Having two telegraph lines would give you redundancy. It would also give you greater throughput if you put both into usage at the same time.

The same redundancy is necessary with data stored on hard drives. Perhaps all your family photos are on a single hard drive. What if that hard drive fails? Doing backups to make copies is necessary. Any application data must never be at risk of being lost or being unreachable. Therefore, some form of data redundancy is necessary with MongoDB.

## MongoDB replica set

If you have a single machine for your MongoDB database, you could configure a MongoDB single-node. Then when that goes down, you cannot access your database until it comes up again. This would be fine for occasional usage scenarios or for development experimentation.

MongoDB has the ability to configure what is called a replica set. This gives you multiple parallel, redundant copies of all data. This ensures that your data will be safe and available. This is what you always get by default through the Atlas portal if you utilize that PaaS.

In a replica set, you have multiple duplicated databases. Only one machine is designated as the primary at any given time. If the primary database server goes offline, a secondary server will take over. The following diagram gives you a general idea of what this looks like:

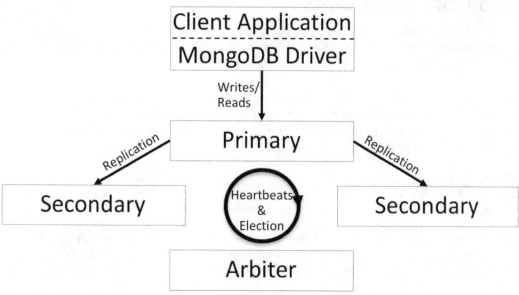

*Figure 18 - MongoDB replica set*

The configuration will look slightly different based on which plan you select from Atlas. The basic idea is that the primary database server receives and fulfills all read and write requests.

All the while, the secondary database servers are kept up to date with all changes. Each database server can be on its own dedicated AWS EC2 virtual machine in different availability zones.

Each server is constantly being checked with a heartbeat signal. If the primary database goes down, the two secondary servers and the arbiter would detect that, and one of the secondary servers would be switched over to become the primary database server. The arbiter is really just there to break any tie votes if needed.

A replica set allows for faster reading of data because multiple copies exist and data can be fetched in parallel from each replica copy, if that is what you want. You must designate reads to be fulfilled by the secondary servers if you determine that is justified. Just be aware that you could get stale data that has not yet been updated by a replication process. This is great, because you keep adding secondary machines and can achieve better read performance now.

*Note*: *With a PaaS solution, you generally do not have control over the replica set configuration unless you work with the provider to get something customized. There are pre-determined configurations you select when you purchase a plan. Nothing is preventing you from implementing an IaaS solution and setting up your own virtual machines and replica set configuration if that would work better for you.*

## Secondary consistency

There is a complication to be aware of in having replications available. Any write to the primary storage collection must eventually make it to all the copies. Therefore, you must make a choice as to how that replication is accomplished.

MongoDB has a setting called "write concern" that allows you to specify if you want a majority of replicas to report that the write has taken place before it is acknowledged or failed. You don't have to require this. If you don't, then all writes eventually make their way asynchronously to all replica database servers.

# 5.3 Sharding

The replication previously discussed stores the same data on multiple machines to provide emergency backup to ensure availability. Sharding also spreads data out across machines. With sharding, a given document appears in only one replica set of a sharded cluster, but would be in the primary and secondary machines of that shard.

The purpose of sharding is to allow you to grow the amount of data you can store and also increase the performance of operations. Both concepts of replication and sharding can be applied at the same time in an architecture. Sharding is just the increasing of the number of replica sets that you have as individual units.

Sharding spreads the data across multiple replica sets. The multiple replica sets in a sharded cluster act as if they were one single collection. The sharding technology knows where to go for any given read or update to make it easy for you to use.

Sharding helps when you have large datasets and are wanting to maintain high throughput. For example, you might have a lot of data constantly being accessed. This can become a bottleneck with a single SSD. If you distribute the load across multiple SSDs, then the CRUD operations would not conflict as much.

MongoDB can be set up to take care of everything for you. You can select a plan from Atlas that has it all set up for you.

Atlas has various plans you can choose from, depending on how much you are budgeting to spend. With the free-tier plan, there is a hard limit with one single SSD block storage for your database, so you can only go up to a certain size and then you can't grow beyond that. Atlas currently only offers replica set and sharded plans (multiple replica sets), and you can't just have a single MongoDB service on its own.

Cluster plans go up to a certain amount of storage. However, working with mLab or Atlas support people, you can keep increasing the horizontal scaling of the sharding by adding more storage. Additional replica sets can, in theory, be added to accommodate your largest data storage needs.

## Reasons for sharding

The concept of data sharding (also called partitioning) was invented to help approach the ideal of being able to store an "infinite" amount of data and retrieve any part of it in a minimal amount of time. Let's dig a little deeper into the scenarios that will cause you to implement a strategy for sharding. Here are some reasons to implement data sharding:

- **Running out of room:** With a limit to storage for a single database SSD, you might simply outgrow that capacity.

- **Machine performance:** There are utilization limits for RAM, CPU and SSD access that might be reached.

*Note*: *With an Atlas PaaS databases, you have the ability to choose a configuration with sharding already configured for you. In the case of mLab and Atlas, you can pick a preconfigured machine architecture and then set up how your sharding will act. If you want to go the IaaS route, then you must configure this yourself.*

## How sharding works

Here is how sharding works. Imagine that you start out with a single MongoDB database server and on that server, you have a single collection. Each document you create could have a property that has a random capital letter chosen from A through Z. You might also set up an index on the letter property. A JSON document you might want to insert could look as follows:

```
{
   "letter": "G"
}
```

At this point, no matter what the letter property value is, all documents will be created in the same database collection. The box below represents a single replica set (primary and two secondary machines). This example shows what this would look like:

*Figure 19 - Collection in a single replica set*

Then, at some point, you realize that you need to add a whole lot more documents and want to achieve a higher level of throughput on your read and write access. The above single-node configuration can then be made into what is called a multi-node sharded cluster.

MongoDB will start balancing documents between the available shards (replica sets) in the cluster to create a more evenly distributed storage. It actually does this in chunks. With additions and deletions of documents happening, MongoDB keeps it all balanced based on the sharding key. You can choose either a hash or a range strategy for you sharding.

Your documents end up being distributed over the three shards in the cluster. See the following figure for a visualization of the distribution using a range sharding strategy:

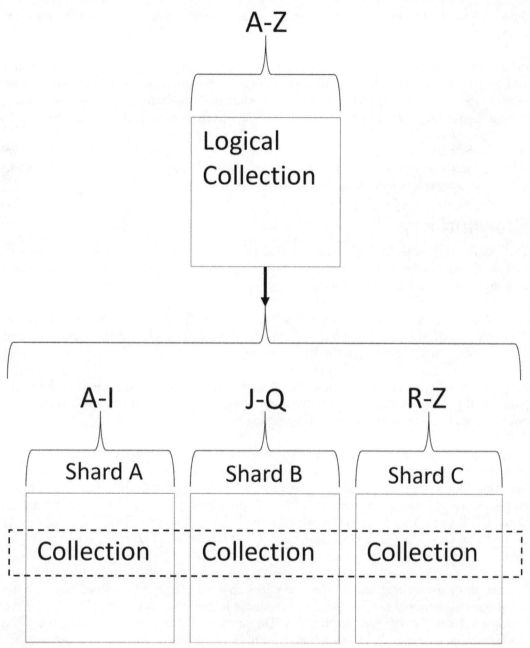

*Figure 20 - Collection distributed over three shards*

As it turns out, each shard is a replica set. When a database request comes into the cluster, MongoDB does all the work to route the request to the proper replica set shard. Your code is

PART I: The Data Layer (MongoDB)

shielded from the fact that this is going on. A single logical collection does all the work for you to coordinate across the actual shards that have the real collections.

I won't go into the architectural diagram showing the components to set this up, but you can look it up online if you really want to implement an IaaS configuration on your own instead of using the PaaS solution. When you use the Atlas PaaS solution, you would most likely enlist a support engineer to help you if you wanted to customize your sharded cluster.

*Note: There is a way to take shards out of your cluster. There is a mechanism to let MongoDB know that this is your intention. Once you do so, MongoDB begins migrating data off of the soon to be decommissioned shard. Once that is done, that shard can be freed up.*

# Sharding key

Your shard must be set up with what is called a sharding key. A sharding key is similar to how an index is set up. With an index, you specify a property that you want to use for a speedy lookup, using some determined algorithm, such as a range or a hash search.

Look at the previous figure, and you will see three shards. The sharding key, in this case, is the property that MongoDB will use to determine what shard each document exists in. For this example, it would have been the letter property.

Each document can only exist in one single shard. A shard key is thus used as a sorting property. If I created a document with the letter property set to 'M', it could be stored in the middle shard because of the range strategy setting it there.

There is a fair amount of thought needed to select the proper sharding strategy and select a property to key off of. Just remember that you must know what your queries are going to look like. Don't forget that you might even have queries that cross shards, like those using range criteria. Imagine if you want documents from the prior example that had a letter greater than D and less than L? The shard service would actually know it needs to send the query to both the first and second shards and then your code would process all of the result set for what you want.

Indexes still exist on each shard. The query lookup would first go to a shard and then the shard replica set would use any applicable indexes to find the document(s). You can have multiple indexes, but only one sharding key. The sharding key must be the same as one of the indexes. In our example, there was an index for the letter property, and that was also used for the sharding key.

If you had documents representing customers, you could look people up by their last name. You could then use a hashed sharding key. That way, queries can narrow the location to one single shard and then quickly retrieve documents from that shard using the index. A hash

## Chapter 5: Managing Availability and Performance

shard key is nice because it can most likely give you a more uniform distribution of documents across shards for a fairly even retrieval cost. This is great for locating documents with a specific query that can zero in on the document.

Range queries might not be as effective with sharding. If you do know you have a good distribution of range values, then perhaps a range strategy would be best. Range sharding is efficient if you have queries where reads target documents within a contiguous range of values.

You also must consider what your queries will look like and what your document composition will look like. For example, you will certainly have a performance problem with hash strategy sharding if you try and do a range type of query that causes all the shards to be searched.

If your query does not actually utilize the shard key property, then the service has no choice but to send the query to all shards in the cluster and then collect all of the results. But at least, you should have considered what the index should be for making that effective.

Like an index, a shard key can consist of multiple property names. You could thus use a compound key such as last name, first name, and city. Shard keys cannot be created for a property that is an array.

You might have been thinking that if you kept adding documents with the same or similar key value that they could all go to the same shard and then the disk for that would run out of memory. This is actually not the case. There is a process going on in the background, regardless of the sharding type (range or hash) that moves data around between shards in chunks. This is called the Balancer. You don't actually set the shard boundaries yourself. MongoDB figures that out for you. Figure 20 was just a fictitious illustration of a possible balancing that could happen.

As was mentioned, you are ultimately limited by the disk space available and that is why you would keep adding shards as you reach the limit of document storage. As a new shard is added, the Balancer does the work to spread documents evenly out across all available shards.

*Note: Let me make sure you have all the terminology down. The Atlas portal lets you create a cluster. This is the set of machines that hold your databases. In their terminology, **a cluster** can be either a **replica set** (primary and secondary machines), or a **sharded cluster** (multiple replica sets that are each called a shard)*

# Chapter 6: NewsWatcher App Development

This chapter takes some of the concepts that you have learned and applies them to a project using the Atlas PaaS offering, hosted in AWS, to create a data layer. You will learn how to get the data layer up and running and learn some best practices along the way.

In this chapter, you will go to the Atlas portal and create the MongoDB database and collection resources for the NewsWatcher sample application. To get started you must first have an active Atlas account.

*Note*: *MongoDB is an open-source project and you could download it for free and run it on any machine you like. This is not the approach taken in this book. You can certainly investigate that option if it better meets your needs. There are other MongoDB PaaS hosting options out there besides Atlas, so do your research.*

You will be setting up the following resources:

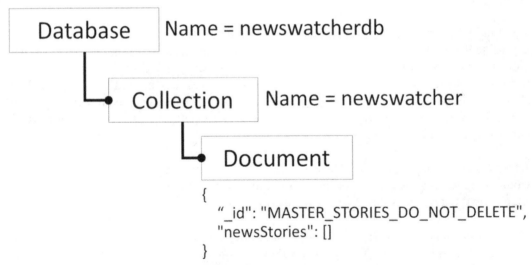

Figure 21 - NewsWatcher MongoDB resources

The only document you will add to the collection right now is the one you should add manually. It is required for the functionality of the NewsWatcher app. You should also manually add a few other documents just for testing purposes to try out a few queries. Later, you will see how documents will be added through JavaScript code in your Node.js process.

Chapter 6: NewsWatcher App Development

# 6.1 Create the Database and Collection

The first task will be to create the database. For the sample application, you can select the option that will give you free hosting. This will be fine for your development and testing purposes until such time that you need to scale for greater storage and performance.

You can also study the other configuration offerings available through MongoDB Inc. You can even try them out for a day or two, as you are only charged for the time you have them available, and you can easily delete them when you no longer want the charge.

It is certainly worth the cost to try out some of the other configurations that allow for other capabilities such as sharding. You might want to take some time to look through the plans and pricing pages on the MongoDB Atlas site to familiarize yourself with what is possible. For example, the amount of RAM and storage and IOPs changes per the different plans. Thus, it is very important to understand your usage needs to be able to select a plan.

When you sign up, you can select the hosting provider such as AWS, the hosting location, and the specifics about the configuration that determine the charge. To create a database cluster for the NewsWatcher app with the Atlas portal for free hosting, you can go to the MongoDB Inc. website (https://www.mongodb.com/). It is fairly self-explanatory from there. First, create a project in your account. Then look for the **Build a New Cluster** button. At some point, you will need to provide a username and password.

After creating your database cluster everything will be deployed and ready for your usage. Once it is ready you need to set up access on your machine for the Compass desktop tool or the Mongo shell.

1. Click the **Connect** button.
2. Click **ADD CURRENT IP ADDRESS** and give it a name. If your IP address will be constantly changing, select to allow all IP addresses. You will still be secure, as access must be authenticated and authorized.
3. Download the MongoDB Compass tool found in the Atlas portal at the bottom of the page, by clicking **Download Compass**. This will be necessary later for your interactions with the database and collection.

The figure below shows you what the Atlas portal looks like at the time you use it to create a database cluster.

## PART I: The Data Layer (MongoDB)

*Figure 22 - Create Cluster page, MongoDB Inc. portal*

In a few moments, you will be all set up and will be ready to start using your free MongoDB database from MongoDB Inc., hosted on AWS EC2 machines. If you look in the MongoDB Atlas portal, you will see your cluster now shows up.

# Set up a connection with Compass

With the Compass tool installed, you will need to make an initial connection to your MongoDB hosted database. Look back at the information you saw in the Atlas portal when you clicked **CONNECT** for your cluster. In there you will find the hostname of your cluster. You can configure Compass to be able to make a connection as follows:

1. In the Atlas portal, click the **CONNECT** button again and then click the area that says **Connect with MongoDB Compass**.
2. Click the Copy button.
3. Open the Compass tool.
4. Click to create a new connection. Everything should auto-populate, and you should be good to go by clicking **CONNECT**. If not, follow the next instructions.
5. Paste in the **Hostname**.
6. The port should already be set to 27017.
7. Set **Authentication** to **Username / Password**.
8. Enter the admin username and password.
9. Leave **SSL** to **Use System CA / Atlas Deployment**.
10. Leave **SSH Tunnel** to **Off**.
11. Enter something to remember this connection by for the **Favorite Name**.

# Chapter 6: NewsWatcher App Development

12. Click **CONNECT**.

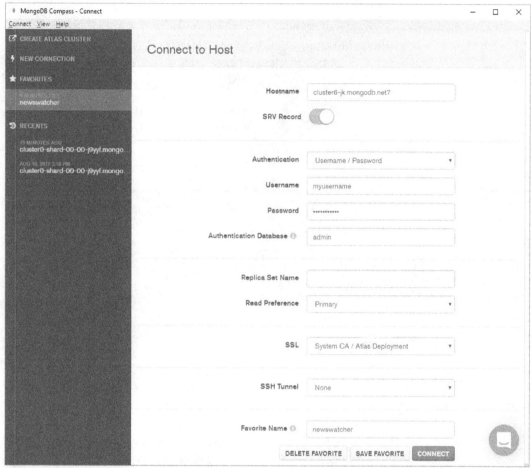

*Figure 23 - Compass connection UI*

# Add a database and collection through Compass

You need to create a database with a collection inside of your cluster as follows:
1. Open the MongoDB Compass app and connect.
2. Click **CREATE DATABASE** (hover over the "+" plus sign at the bottom of the left pane) and type in the **Database Name**. I entered "newswatcherdb". Enter a name for **Collection Name**. I entered "newswatcher".
3. Click **CREATE DATABASE** in the form.

*Note: If Compass will not allow you to perform edits, it is probably because you are connected to a secondary, and not the primary machine in the cluster.*

79

## PART I: The Data Layer (MongoDB)

You can now create the one required document that must be manually created. You will see later how this document fits into your data model. To create the document, do the following:

1. Click on the newswatcherdb database.
2. On the newswatcher collection, click **INSERT DOCUMENT**.
3. Type in the document content as shown below. Make sure to select a data type of Array for the newsStory property. Click **INSERT**:

```
{
    "_id": "MASTER_STORIES_DO_NOT_DELETE",
    "newsStories": [],
    "homeNewsStories": []
}
```

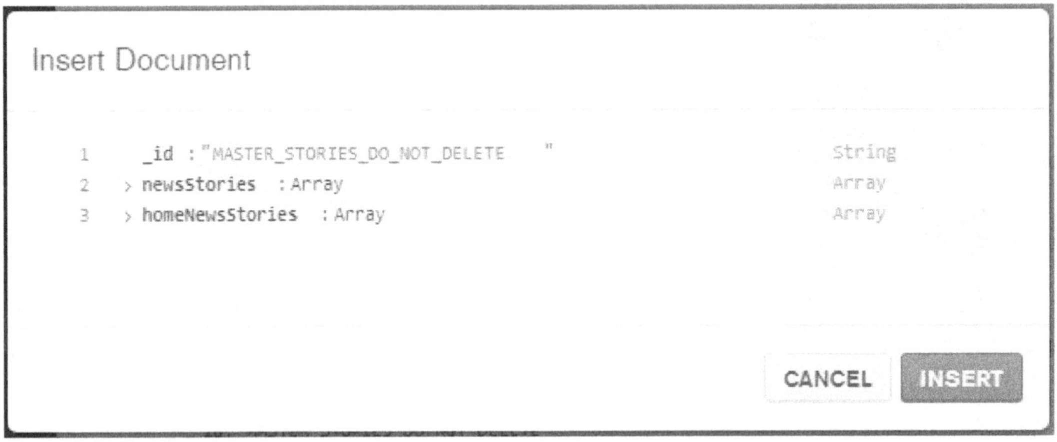

*Figure 24 - Create a document*

To add a new property/field, you can hover over a number on the left and it will change to a plus sign and then you can add one. You also need to change the type with the drop down you will find there. After clicking **INSERT**, you will see the document on the collection page. Isn't PaaS wonderful? There is no setup or maintenance or worrying about if you have the latest version of MongoDB.

## Install of the mongo shell

The Compass tool does not now allow for easy creation or import of documents, because you have to create them a property at a time. To import a larger document easily, you need to install the mongo shell. In the Atlas portal UI, you will find a link to download the mongo shell. You will want to select a custom install with the selections looking as follows (see figure 25). Otherwise, you will get the database installed locally along with the mongo shell. You don't actually need the mongo shell to go forward developing the NewsWatcher application, so don't install it if you are hesitant about it.

Chapter 6: NewsWatcher App Development

On the UI page of Atlas that shows the connection information, you will find a string to copy that gives you the information necessary to make a connection. Give the full path to the mongo executable and add the connection string to that. It looks as follows to run the mongo shell:

```
"C:\Program Files\MongoDB\Server\3.4\bin\mongo.exe" "mongodb://cluster0-shard-52-78-k6yhs.mongodb.net:27017, cluster0-shard-52-78-k6yhs.mongodb.net:27017, cluster0-shard-52-78-k6yhs.mongodb.net:27017/test?replicaSet=Cluster0-shard-0" --authenticationDatabase admin --ssl --username max --password blah
```

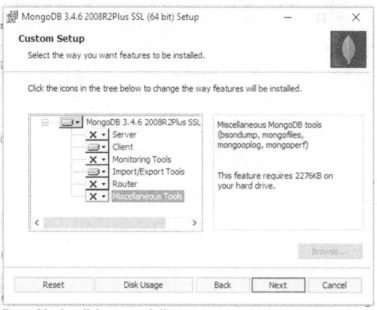

*Figure 25 – Install the mongo shell*

## 6.2 Data Model Document Design

It is time to diagram out the structure and relationships of the document types that you will need for the NewsWatcher application. This is definitely an iterative process where refinements are made over and over until it is correct. Even after you have implemented a data model, you may find that it does not give you the performance you expected, and you might end up altering the design.

Think again about what the requirements are for the NewsWatcher application and you can understand what is needed. NewsWatcher will have users that log in. Thus, you have identified that there is a need for a user document.

# PART I: The Data Layer (MongoDB)

There is also a single document that holds the master list of news stories. There will be some code that is run every few hours to collect news stories and store them in that document.

A third document type is for the news stories that users share and comment on. There would be multiple User and SharedStory documents, but only one MajorStories document. This model is completely denormalized, so there are no keys to link any documents together and there will not be a need for any type of join operations. The following diagram shows some of the needed documents:

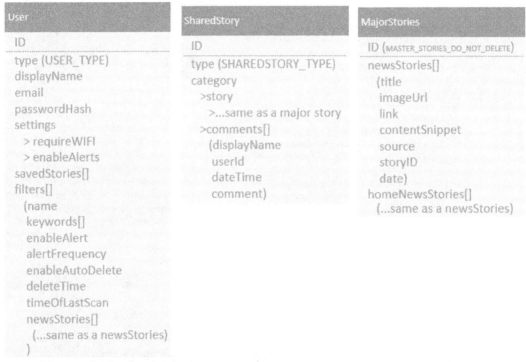

*Figure 26 - NewsWatcher documents*

Let's look at what the User document contains. In there, you will want to include an email address for each user. This will uniquely identify your users and allows them to sign in. Users must also enter a password. You can safely store a hashed value of the password (you should never store a password in plain text). Then you can let users pick a display name that other users will see when a user comments on a shared story. You should never reveal their email to anyone else.

Next, there should be certain global values that can be used for user preference settings. You can put that in a sub-hierarchy called `settings`. For example, you might want to give users the option of not using any cell phone data and restrict the app to using Wi-Fi only.

# Chapter 6: NewsWatcher App Development

You can assume that there will be some users that would like an alert feature for when news stories come in to be immediately notified. You could create a Boolean value for that.

The compelling feature of NewsWatcher is the ability to have the app scan for the news a user cares about. NewsWatcher users are not the type that wants to go to some general overall curated news page, but are interested in customizing their own specific filtered view of their news. This is done by filtering news stories with keywords.

Users can set up as many filters as they like, so you can conclude that the design requires an array of filters. Each filter will need to contain a title for the filter, keywords, time of the last news scan and a list of stories and their time of capture. The list of stories for a filter is populated by scanning the master story document to see if there are any matches with the keywords.

NewsWatcher has the ability to save off interesting stories per each user so that they appear separately. This is what the `savedStories` property is used for. We won't actually implement that at this point.

The other properties shown in the user document are for other features as outlined in the requirements. Those other features won't be implemented though.

This will give you a good start at an MVP (Minimum Viable Product) to go out with. If you do a bit of advanced thinking, you can model all of this in your diagram and just not implement everything yet. You can feel confident that your data model can accommodate your future needs.

## Entering some test data

At this point, you can open the page for the newswatcher collection and add a document to the collection for testing things out. You can of course only do this if you had done the work to install the Mongo shell. You can insert documents with Compass, it will just take you longer.

1. Start the mongo shell as shown in section 6.1.
2. At the shell prompt, you can type "**use newswatcherdb**" to direct commands to that database.
3. Run the following command to insert a document:
   ```
   db.newswatcher.insertOne(
   {
     "type": "USER_TYPE",
     "displayName": "Bushman",
     "email": "nb@hotmail.com",
     "passwordHash": "XXXX",
     "date": 1449027434557,
   ```

## PART I: The Data Layer (MongoDB)

```
      "settings": {
        "requireWIFI": true,
        "enableAlerts": false
      },
      "savedStories": [],
      "filters": [
        {
          "name": "Technology Companies",
          "keyWords": [
            "Apple",
            "Microsoft",
            "IBM",
            "Amazon",
            "Google",
            "Intel"
          ],
          "enableAlert": false,
          "alertFrequency": 0,
          "enableAutoDelete": false,
          "deleteTime": 0,
          "timeOfLastScan": 0,
          "newsStories": []
        }
      ]
    })
```

4. Run "**db.newswatcher.find()**" at the prompt to see any documents in the collection.
5. Run "**exit**" to quit the mongo shell.

You will see the document added. It has an automatically assigned `_id` created. You can launch the Compass application and view your created documents and delete or edit them as necessary.

For now, you can go ahead and experiment by creating a few more User documents in this same collection. Later, documents will only be added through code. At this point, all you are interested in, is being able to test out some queries before developing the next layer of the application. You can get a feel for how the portal UI is used and learn about how queries are constructed before you put those into code.

If you read through the MongoDB documentation you will find that there are ways to do bulk importing or exporting of documents. For example, there are tools like mongoimport and mongoexport that you can run from the command line. For example, an export might look as follows:

```
mongoexport -d test -c records -q '{ date: { $lte: new ISODate("2017-09-01") } }' --out exportdir/myRecords.json
```

# 6.3 Trying Out Some Queries

You might have entered a few documents by hand in a collection. You can now try out some queries against that data through the Compass application or the Mongo shell. At this point, you just want to get a feel for what the tool looks like and to be ready to learn about how queries are constructed before you put those into the service layer code.

You will use the same UI shown in earlier chapters to run queries against your MongoDB collection. That is where you utilize the criteria syntax to query and specify what you desire to see in the output.

Try some queries like the following:

```
{"type": "USER_TYPE"}
{"type": "USER_TYPE", "email": "nb@hotmail.com"}
```

Now set the **PROJECT** in the options area of Compass to be `{"displayName": 1}` and try the query again.

If your query syntax is incorrect, you will be notified of the error. However, if you mistype the name of a property you want to project or query for, you will not get an error but will get an empty result instead. For example, try the projection criteria property name as `{"blah": 1}`. If you do this, you will not get an error but will get an empty result set.

Keep in mind that MongoDB is a schema-less database and it assumes that the "blah" property could be there in the future, but it just is not there right now. Properties can come and go in a schema-less document-based database.

# 6.4 Indexing Policy

You could write code that uses the MongoDB API that runs to create your needed indexes. My approach is to not put things in the code that are one-time configurations. I will instead prefer to use the Compass application to create indexes. You could also use the mongo shell to run a command to create the index.

For the NewsWatcher application, you can imagine that you would have a query in the service layer that will look up a user by their email. You would want to add a specific index for that by doing the following:

# PART I: The Data Layer (MongoDB)

1. Open the Compass application, click the newswatchedb database.
2. Click the newswatcher collection.
3. Click the **INDEXES** tab then click **CREATE INDEX**.
4. Give the index a name.
5. Select "email" as the property and "1 (asc)".
6. Select the checkbox for **Create unique index**.
7. Select the check box for **Partial Filter Expression** and enter "{email: { $exists: true}}"
8. Click **Create**.

You check **Create unique index** as you don't want email addresses duplicated across users. This is a way to uniquely identify an account for a person. You have also set up what can be called a "sparse index" using the Partial Filter Expression. This means that documents that don't have the "email" property will not be used in the index. This will give you lower storage requirements and offer better performance with the index maintenance.

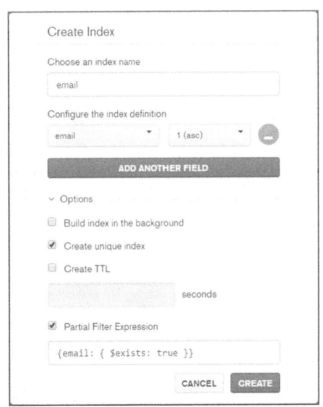

*Figure 27 - Creating a new index*

## 6.5 Moving On

This completes the work to get the data layer up and running. You can see that it was all about setting up your configuration through the Atlas management portal and the Compass app. You did not need to write any code yet. This means that you are postponing the writing of any Node.js JavaScript server-side functionality until you work on the service layer.

### Testing

For the NewsWatcher application, the service layer is actually the proving ground for the data layer. The service layer will connect directly to the MongoDB collection and perform CRUD operations. There will be functional tests put in place to prove that the data model works. You will also be able to take care of the nuances that go along with the data layer, such as performance tuning and concurrency issues.

In reality, the best way to develop software is to work on it in terms of vertical slices of functionality. This means that for any features you have thought up, you would implement it in all three architectural layers at once.

# Chapter 7: DevOps for MongoDB

In this chapter, I will go over some of the operational responsibilities for managing a MongoDB database. For example, with NewsWatcher, you know that the data layer stores user accounts and news stories. You can think about what concerns you would have with that. Daily DevOps work will involve the monitoring of the database. You can take a look at what the Atlas management portal will let you monitor.

You will want to make sure that data access is secure and performant. You can set up replication, sharding, and indexes. Once those are set, you should leave them alone until some change comes along that causes you to tweak them for a specific reason.

## 7.1 Monitoring through the Atlas Management Portal

One view you can look at is the view of machines in your database replica set. Here is a screenshot that shows all the machines in a replica set (primary and two secondaries):

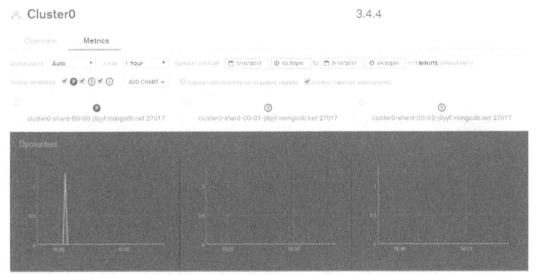

*Figure 28 - Atlas management portal server view*

Chapter 7: DevOps for MongoDB

*Note*: *The free tier offering is limited in what it offers in the Atlas portal. For example, there is a Data Explorer with the paid tier. The Metrics page for a paid tier has a lot more information available like - Sharded Cluster Metrics, Replica Set Metrics (more metrics), Real-Time Tab, Status Tab, Hardware Tab, DB Stats Tab and Chart Controls.*

## Telemetry charts

You can drill down further into the performance metrics of each of these machines by clicking on them. There are charts provided in the Atlas management portal to show you the server utilization numbers. The real-time telemetry values will let you know things like how much storage you have used up.

Here is an image showing the Atlas management portal monitoring page with the telemetry that is shown by default. In a replica set, you will have more than one machine, so you have to pick the primary or one of the secondary machines if you want to see their telemetry separately.

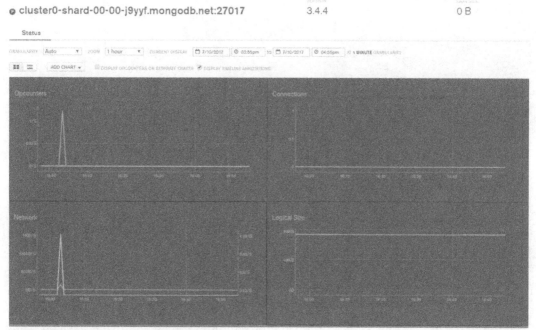

Figure 29 - Atlas management portal metrics page

## Telemetry alerting

You probably want to be aware of how much storage is left for your database. You could set up an alert to notify you when you are reaching this limit. You might also be concerned about machine performance and set up some alerts around specific performance measurements. If you see your machine performance degrading, then you can shift to a more capable configuration with the PaaS offerings of MongoDB Inc.

If you go into your Atlas portal, you can select **Alerts** from the menu on the left. Then you can select the appropriate tab from there. For example, there is a tab to view and acknowledge alerts that have been triggered.

Click on the **Alert Settings** tab to see what alerts you have by default and add any you need. In the figure below, you can see that there is an alert that triggers when you have reached 90% of your storage capacity.

*Figure 30 - Atlas Alert Configuration page*

You can select from a wide choice of possibilities for how you get notified of an alert being triggered. Click on the **Add** button to see these. The selections include – Atlas User, Email, SMS, HipChat, Slack, Flowdock, PagerDuty, and Datadog.

Chapter 7: DevOps for MongoDB

# 7.2 The Blame Game

Once your data is secure and has been tuned for the best performance, you really don't need much in the way of day-to-day care anymore. Believe me though when I say this – your potential troubles are not over by any means. From my experience, you will be spending your time caring for the integrity of the data as much as anything else. This is especially true if there are a lot of other systems integrating with yours that touch the data at some point.

Unfortunately, every time someone sees a data corruption problem, they come to blame whoever is in charge of the DBMS. You will hopefully have confidence that most of the time the accusations are unwarranted, and you will be able to track the problem down to some supporting system. For example, some external system that is feeding you data may suddenly have missing, intentionally altered, or corrupt data. It is a good idea to put data validation measures into place at all the points of integration.

You would also be wise to put some handy scripts in place to allow you to diagnose issues and fix them. For example, you might need to recover data from a backup snapshot or reimport data in bulk from a dependent system.

# 7.3 Backup and Recovery

The good news is that you don't have to worry about disaster recovery. The bad news is that you have to worry about disaster recovery. It all depends on what your definition is of a "disaster".

With a replica set in place, MongoDB stores multiple copies of your data that are always in sync with the primary server. This means that, within a region, you have redundancy in case of network or drive failures. This is the case with the different AWS availability zones that each EC2 server is in for your replica set. You have this set up for you through the PaaS plan you select and don't necessarily have to deal with it directly yourself.

This replication means that your data is safe from drive failures, machine reboots, power outages, network outages, and such. If the drive that the primary copy of your MongoDB database is on goes bad, you are covered. MongoDB and AWS will take care of rotating this drive out and moving you to a new primary drive and adding a new replacement backup.

## You still need backups

Data replication is not the same as performing a data backup. Just because you have replication does not mean you are protected from somehow losing or mangling your own data by mistake. It is a good idea to institute a backup process to periodically store a snapshot of all your data. That way you are able to recover from inadvertent corruption or loss of your data.

Backups are useful in many scenarios. For example, you might have some bug that was introduced in your code that causes all your data to get corrupted. You then need to roll back to the database copy you had before the data was corrupted.

You could, for example, write some code to copy data from one MongoDB region into another region and keep that as a backup.

You could also save a collection as a file for safe-keeping on some local machine you have, or place it in EBS or S3 storage in a compressed form. Then you can do a restore programmatically or use a tool to import everything from your snapshot. You can export a JSON file and store it yourself if you don't want to spend the money on database collections or other online storage being used for backups.

The Atlas management portal has a **Backup** selection on the left-hand side that lets you create an immediate backup, or to schedule a time each day for one to automatically happen. There is a pre-determined retention policy for each specific time-related snapshot. The core MongoDB project has backup utilities you could also use to perform data backups.

# PART II: The Service Layer (Node.js)

Part two of this book will teach you what a service layer is. You will create an HTTP/Rest API that interfaces to the data layer. This will set things up to be prepared for the development of the presentation layer application that talks to the service layer. Node.js/Express.js and JavaScript are the technologies of choice for this service layer.

There are many decisions that go into creating a service layer. The first thing to design, is what type of interface is needed over the data. This involves separating out the different types of data that your REST interface will expose. This requires you to think about the JSON payloads that get transferred back and forth for each request.

You will learn how to use a Node module to call into the MongoDB data layer developed in the first part of this book. At the end of this part, the application will be fully functional and ready to integrate with the presentation layer.

This book will be using Amazon Web Services (AWS) to host the Node.js application.

The extremely important topic of testing will be covered, and you will learn how to use the Mocha test framework to run your tests. You will learn about functional as well as performance load testing.

In order to give full coverage to the topic, I will also discuss what it means to set up all aspects of the day-to-day operations of Node.js for actual data center operations management. You will learn how to manage a PaaS environment and how to do debugging of issues. Security will be an important topic that is also covered.

*Note*: Many people refer to Node.js simply as Node, and I often do the same.

# Chapter 8: Fundamentals

This chapter presents the fundamental concepts of the middle-tier of the three-tier application architecture that is being outlined for you in this book. You will learn what the middle-tier is typically composed of. I can then get into the specifics and show you how Node.js can serve as a middle-tier service layer.

## 8.1 Definition of the Service Layer

The service layer provides the core capabilities of a three-tier architecture. The whole idea of a service layer is to build an abstraction layer over business logic and data access.

If you really simplify down the concepts of a three-tier architecture, you can say this about the lower and upper layers - the lower data layer just stores data and the upper presentation layer just displays the User Interface. That leaves the middle-tier service layer to do all the rest of the work. In most applications, you will certainly find more code in the service layer than in the other layers. The following diagram shows this simple view:

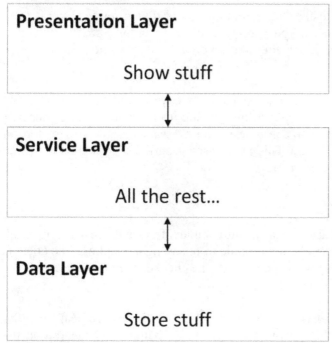

*Figure 31 - Simplified three-tier architecture diagram*

It would not be reasonable to have the presentation layer handle the business workflow logic. You also do not want to expose business workflows in the data layer. The data layer should be kept as simple as possible and should only handle the CRUD operations and perhaps more difficult data transaction logic. You will normally find the more complex business services code in the services layer.

*Note: Some architectures split the middle-tier out into a services and a business layer. These two concepts are combined in this book. You will find that all the needed functionality of this layer can be accomplished within the single technology framework of Node.js, along with the use of numerous npm packages.*

# A contract of interaction

A service layer can actually be created with no particular UI in mind. For example, there are major companies that expose their APIs so that anyone can interface with their backend and write their own UI. Companies like eBay, Twitter, and Facebook have been successful at this. For example, you can interact with the eBay API to bid on items.

In the case of the NewsWatcher sample application, there is just one single UI that I wrote, but anyone else could write a different UI on top of my exposed Rest API.

Regardless of whether you are tied to one single front-end UI, or if your service layer is open to allow many applications to connect, you need to think in terms of a strict contract of interaction. This means you must define the connection routes up front and the JSON messages that are required.

I will not be using any specific connection standard like you see with standards such as SOAP. You can explore things like Swagger on your own if you are interested in making your API generally available to people in an easy to consume and formal way. You can also explore an AWS offering known as the API Gateway. This can be used to surface a formal API contract that sits in front of your Node.js service layer.

# An abstraction layer

A service layer is built in a way that abstracts away the complexities that go on in the backend. This will shield the presentation layer client-side code from any tight coupling. The client is also protected from any backend rework. In many cases, the client doesn't even need to make any changes when backend code is rewritten.

It may be, that a single call to the service layer results in a series of backend calls that are each processed in what can be called a workflow. This coordination falls squarely in the service layer in order to hide the complexity. The multiple backend services comprise your overall architecture and can be unified through a single Node.js entry point.

For example, you might have one service that does all of the storage and retrieval of all user account information. Another service might contain billing account information, and yet another might deal with order information. You should never build one single monolithic service that does everything. Take the time to split up your platforms into discreet services. Each of these will serve a role, be self-contained, and operate independently. Research what is called a microservices architecture.

This does not mean that external clients need to authenticate and connect to each of the microservices individually. In other words, you might want your presentation layer to only have one single service layer entry point that coordinates calls to other independent microservices that each exists as autonomous services. You will actually be grateful you have done this, as you can make rather significant changes in the lower layers and minimize the code that has to change in the client.

Your backend might already consist of "legacy" systems that were not written with Node. These backend systems might be written in different languages and be running on your own proprietary, on-premise platforms. In this case, you can decide to write a gateway service layer with Node and have that code interface to your backend systems. Node is great for routing and orchestration. It can work to process requests asynchronously with a high degree of concurrency. AWS offers an API Gateway that you can also investigate.

All or part of your systems can have their own Node.js interfaces. It is up to you to decide what is worth your time and investment to do. The following diagram illustrates the gateway layering that could exist in a services layer:

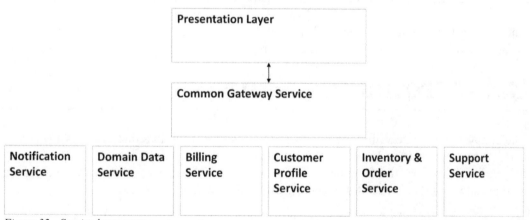

*Figure 32 - Service layer gateway concept*

If you decide you need to replace the billing system, the gateway service provides the abstraction and protects the client from any changes.

## Service layer planning

Knowing what operations and workflows are needed is the first step in creating a service layer. Use the following questions to help you determine the design of your service layer:

- What operations and workflows are needed?
- What are your security and privacy requirements?
- How are people authenticated and authorized?
- Is there a need for a pub/sub notification system to deliver push notifications?
- Is there any type of domain-specific configuration required?
- Is programmatic resource management required? Scripting or templates?
- Are all data interactions encrypted?
- Are multiple systems going to call this? Would a message contract schema be appropriate?
- What validation can be made at the interface to not let anything invalid in?
- What meaningful errors need to be returned?
- Do you need user roles and access control?
- Do you need caching on top of your database layer?
- Is there any periodic processing of the data? Is it periodic in the background or real-time? In batches?
- Will you need to queue up work and have workers process work asynchronously?
- What are the SLA requirements for all operations?
- What are the access volume and the rates per time period?
- Will there be bursts of activity or is access evenly distributed?

The answers to these questions should be carefully considered. Consult experts along the way before you roll anything out into a full production environment.

# 8.2 Introducing Node.js

The simplest way to describe Node.js, is to state that it is a runtime that executes JavaScript code that you provide it. You might think I just described a browser environment for you. After all, this is what the Chrome browser can do. The Chrome browser has a JavaScript engine called V8 that is used for executing client-side JavaScript code.

Node.js, however, is more typically running on the server-side. To accomplish this, someone (Ryan Dahl) actually took the same V8 engine mentioned, and made use of it in a server-side process and then added additional functionality that would make sense to have running in server code.

# Chapter 8: Fundamentals

You can think of Node as an abstraction layer on top of your operating system so as to make your code run as platform independent code and get many of the capabilities that an operating system provides (file system, processes, network etc.). Node.js provides the overall runtime execution environment.

## Platform independence

Let me put it this way - you could go and write an application in C++ for Windows that implements a web service that listens to and responds to requests and interacts with the file system. But what would you have to do to take that application and make it run on Linux? You would have to port it of course, which requires a re-write of that C++ code to use operating system libraries found on Linux machines!

To get that to run on a Linux OS, you would port your code to use those system calls that are available on Linux. This image shows that you would be writing your code over and over for each platform.

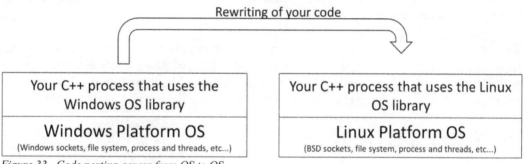

*Figure 33 - Code porting across from OS to OS*

Node lets you write your application code once, and then Node handles the lower level porting of Operating System level calls for you. Node.js acts to abstract away the platform OS capabilities. Not only that, but Node allows you to write all your code in JavaScript!

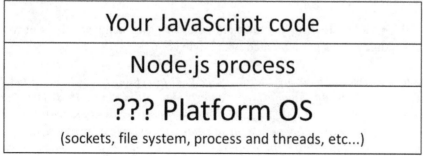

*Figure 34 - Writing code once*

99

PART II: The Service Layer (Node.js)

Node.js is an open-source project and has been ported to run on many different operating systems. Much of the core code of Node.js is written in C/C++ to enable native integration with underlying operating systems and achieve the fastest possible performance. It also utilizes the Google V8 engine to execute JavaScript. V8 actually compiles client JavaScript to native machine code, such as for x86 machine architectures, for faster execution.

Node.js is well-suited for network-based I/O applications. Using Node, it is extremely simple to piece together a web server similar to IIS or Apache. You can easily set up a web service to expose an HTTP/Rest API that works with JSON payloads. Node really fulfills a lot of purposes that allow it to satisfy all the requirements of a middle-tier service layer.

## Extensibility of Node

The real power of Node comes through its extensibility. Node was written to provide the core runtime of execution, scheduling and notification capabilities. Its functionality is then greatly increased through the many extension modules written for it. For example, there are modules for functionality such as WebSockets, data caching, database accessing, asynchronous processing, authentication, and many others.

Node is widely adopted, and people are constantly improving it and creating new modules for it. There is a large community of developers that has generated many extremely useful modules to give rich functionality to your application. Since the patterns of using these modules are all very similar, it is extremely easy to consume them without any learning curve.

## JavaScript bliss

Since Node.js executes JavaScript, there is a consistency throughout the application stack that you are building. JavaScript design patterns are utilized in the construction of modules that become the self-contained components that are available to consume. You can download and integrate those offered by others, as well as create your own.

There are also unit-testing frameworks that work with the JavaScript language that you will use to test your Node.js service layer.

JavaScript is now utilized as more than just a client-side scripting language. It is now a formidable server-side language, as implemented with Node.js.

*Note*: Lest I get a slew of emails accusing me of living in a fairytale, I will make a brief comment on the sensibility of using JavaScript in enterprise applications. It is true that JavaScript can be a challenge with respect to delivering on quality. However, with proper design patterns and testing, you can today create large enterprises services of superb quality with Node.js and JavaScript. Many large companies have already done so. Perhaps as the JavaScript language evolves more features will convince the skeptics that it is here to stay.

# 8.3 Basic Concepts of Programming Node

A simple Node program that you can write would be a program that outputs text to the terminal process window. The following is an example of what this would look like:

```
console.log("Hello World");
```

You can type this into a file, save it to disk, and then have the Node process execute it. If you had Node installed, you could open a command prompt and type the following, substituting the name of your file for *<filename>*.

```
node <filename>.js
```

If you haven't already done so, you should go ahead and install Node on your machine at this time. Go to https://nodejs.org/. You will see a download link labeled "LTS" and another labeled "Current". You want the LTS version (Long Term Support), as that is the stable one. The other one is from the latest code under development and has not been sufficiently proven.

Go ahead and create a file named server.js, place the `console.log("Hello World");` line in it and run node as shown with this file as an argument. You need to be in the same directory as your file.

You will also notice that the process does not stay running. In this case, once your code in your file has been executed, the Node process exits. This does not have to be the case. I will soon explain what would cause a Node application to keep running so it can continually perform server-side processing such as the processing of web requests.

## The REPL

Node has several different options to control how it runs. If you were to leave off the JavaScript file, Node would default to what is called REPL mode. REPL is the mode where you get a prompt and can enter JavaScript to be executed as you type it in. REPL stands for Read Evaluate Print Loop and is a common thing for execution frameworks to provide. For example, MongoDB provides something similar, called the Mongo Shell.

In this book, you will not be using the REPL and only need to be concerned with the main means of invoking Node as shown already, passing your JavaScript file as an argument.

## Node executes JavaScript

Here is another file to try. This one illustrates a bit more code that Node can execute. Running this will result in "HI THERE 343" being displayed:

```
var x = 7;
var s = "Hi there";

function blah(num, str) {
    if (num == 0) {
        return "Can't do that";
    }
    return str.toUpperCase() + " " + Math.pow(num,3);
}

var result = blah(x, s);

console.log(result);
```

This illustrates some basic JavaScript language capabilities. If you look at the JavaScript specification, you could see more of what is possible. You have datatypes, operators, structured programming, logic control, built-in objects and much more.

If you have programmed browser scripts before, be careful about what you assume is available in JavaScript. For example, there is a function named `setTimeout()` that you might assume is a part of the core language of JavaScript. This is not the case. This is where browser implementations have added functionality to JavaScript.

The `setTimeout()` function is indeed provided, but it is implemented through the Node.js layers and not the JavaScript language itself. This is also true for the `console.log()` function.

As was mentioned, the Node process will run and exit upon execution of your lines of code. This is because Node.js will only run while it knows it has code to execute. Try running the following code:

```
while(true) {
  console.log("Hello World");
}
```

To stop the Node.js process, you will have to press **<Ctrl> C** on your keyboard, or close the window. Later, you will see some code that requires Node.js to run forever because it has been set up to respond to events that could perpetually happen.

Chapter 8: Fundamentals

## Using built-in modules

As was explained, Node extends the basic capabilities of JavaScript by providing a set of built-in modules. There are quite a few of them, and you can review them if you go to https://nodejs.org/en/docs/. Click on an API version on the left to see the list of modules. You will see things on the list such as HTTP, Net, OS, Crypto, File System, Console, and Timers.

You can now see how to use the Node.js provided functions such as `console.log()` and `setTimeout()`. Here is some sample code that takes advantage of these added capabilities:

```
setTimeout(function () {
  console.log('World!');
}, 1000)

console.log('Hello');
```

If you are not familiar with how `setTimeout()` works, you have to be aware that this schedules a callback function to run some number of milliseconds in the future. Thus, the string `Hello` prints first and then one second later you see `World!`.

Let's now look at how easy it is to create an HTTP Web server with code that Node would execute. The following example is all the code needed if you use the built-in HTTP module:

```
var http = require('http');

var server = http.createServer(function (request, response) {
  response.writeHead(200, {"Content-Type": "text/plain"});
  response.end("Hello World\n");
});

server.listen(3000);
```

If you were to execute this with Node on your local machine, you could then open a browser and navigate to http://localhost:3000/ and see your `Hello World` message appear.

The only non-obvious line of code is the very first one. This is a function defined in Node.js that you call to load the HTTP module that provides the web server capability. Those who write browser scripts in JavaScript use an 'import' statement instead (see Node 10 docs for usage).

Node uses the concept of modules as its extensibility mechanism. The `require()` function simply returns an object with functions and properties placed on it that are specific to that module for you to use. In this case, you call `require('http')` and get an object, and then use the `createServer()` function from that object. Some things like the `setTimeout()` function are made available globally without you needing a `require()` statement.

PART II: The Service Layer (Node.js)

# Using external modules

In some programming languages and runtimes, you mostly rely on what comes built in. For example, this is the case with the combination of C# and .Net. In Node, however, you will constantly be looking for external modules to add to your application to give you many of your capabilities. As a matter of fact, since there are so many of these external modules provided online, you will need to become good at searching for something and then determine what the best option is.

There is a package manager site you can go to for searching and downloading modules. Go to https://www.npmjs.com/ to take a look. Be aware that some of what is found in NPM are code modules you can use, while other downloads are actually tools you can run, such as the testing tool called mocha.

When the NPM modules are utilized, the layers of code now look as follows:

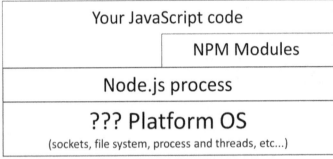

*Figure 35 - Adding NPM modules to the architecture*

This gives you the highest-level view of the different blocks of code.

**Note:** *Regular Node modules are written in JavaScript and distributed for others to use. There is, however, a way to write a module in C++ that is referred to as an add-on. If you do that, you could get better performance (more sustained compute intensive code) and also access to operating system APIs that are native to a machine that might only be available in C++ libraries. N-API is a newer capability provided by Node for building native Add-ons. Learning to create Addons is not necessary for most projects and the topic will not be covered in this book.*

To use external modules from NPM, you need to have them installed alongside your own JavaScript file. There are two steps needed to use an NPM code module:

1. Create a file named package.json.
2. Run the command "`npm install <module> --save`" for each module you want to use.

## Chapter 8: Fundamentals

Running the npm install command will actually add a line in your package.json file. The very first time you run it in a directory, a folder named node_modules is created. If you look in that folder, you will see all of the module code. When you run Node with code that requires one of these modules, Node will be able to find them.

You can try the steps by creating the package.json file with the following lines in it:

```
{
   "name": "test",
   "dependencies": {
   }
}
```

Now run the following command:

```
npm install async --save
```

After running this, a node_modules folder is created, with the async module files installed. Your package.json file will look as follows:

```
{
   "name": "test",
   "dependencies": {
      "async": "^1.5.2"
   }
}
```

There is a convention in the package.json file for listing modules. The name is listed followed by the version you desire. You can specify an exact version, or you can specify just the major number and have NPM get the latest minor version. The '^' character for our usage of the async module means that if any time you refresh your usage with an "npm install" command you would bring in any minor or patch updates. For our example above, that would not bring in a 2.x.x version, as that would be a major version update.

If you add dependency modules in your package.json file by hand and specified the exact version, you could then run the "npm install" command. NPM will install the latest version for what version numbering you specify. The following command is what you run if you edit package.json first and then want to install the modules you specified:

```
npm install
```

It is typical for developers to stick with a known good working version for each of their modules and not update to any new major versions even when they become available. When you feel you need some new capability or security patch of a newer version, then do an update and do extensive testing of everything again. Major versions have new functionality.

# PART II: The Service Layer (Node.js)

Here is another code sample you can run. It makes use of the async module that was installed. Put this code in your server.js file:

```
var async = require('async');
var fs = require('fs');

async.eachSeries(['package.json','server.js'], function(file, callback) {
  console.log('Reading file ' + file);
  fs.readFile(file, 'utf8', function read(err, data) {
    console.log(data);
    callback();
  });
}, function(err){
    if( err ) {
       console.log('A file failed to load');
    } else {
       console.log('All files have been successfully read');
    }
});
```

You will learn more about the async module later. This is basically using the async module capability to sequence through an array of values and do what processing you want on each entry sequentially.

Note that you are using the fs module. You don't need to run an install, or even list the fs module in the package.json file. This is because this module is part of Node.js, but you still need the require statement. If you now run "`node server.js`", you will see the contents of your files printed out.

*Note: Deployment of your Node.js application is easy. You can just copy everything, including the node_modules directory, to a machine. When you use a PaaS environment install, such as with AWS Elastic Beanstalk, you don't need to copy the node_modules directory. AWS will run the npm install for you. If you want version numbers locked, set the specific versions or use what is called a shrinkwrap file or use a feature of NPM 5 for that.*

## Callbacks and concurrent processing

To start with, you need to understand that there is just one main thread of execution in your program. This is the thread that starts up your application and begins execution of your JavaScript code. From there, Node.js sends all your code to the V8 JavaScript VM engine, and OS ported calls to begin its execution.

Everything at the highest level of your JavaScript places processing time on a single thread. Only one thing can happen there at a time, so write code that is not compute intensive.

# Chapter 8: Fundamentals

*Note: VM stands for Virtual Machine and is a concept that V8 uses to isolate JavaScript execution. Don't confuse this definition of a VM with a VM that you find hosted in AWS or Microsoft Azure.*

You have already seen Node code that uses the callback style of coding. This style is prevalent in everything you do in Node. The Node.js library provides for these non-blocking asynchronous callbacks. Your code never blocks, but returns immediately and then at some later time, the callback function is executed. This gives you the concurrent execution capability that Node is famous for.

*Note: There are other mechanisms for doing asynchronous code such as using Promises and Async/Await. These are also very popular and are basically a syntax preference. Async/Await will not be covered in this book.*

## Code execution flow

The code execution path is interesting to trace through. I will now explain a bit of how this all works. Look at the following code that will be used to help understand the execution flow for your JavaScript code:

```
var x = 7;
var s = "Hi there";
var fs = require('fs');

function blah(n, s) {
   if (n == 0) { return "No"; }
   return s.toUpperCase() + " " + Math.pow(n,3);
}

console.log(blah(x, s));
fs.readFile('package.json', 'utf8', function(err, data) {
   console.log(data);
});
```

If this code is in your server.js file and you execute it on the command line as `node server.js` your execution looks as shown in figure 36. You can see the code boundary crossings. You can also see that almost everything is non-blocking. At a lower level, there is a thread in Node that does end up being blocked, but this thread does not affect you at all. Your code still has the callback that is asynchronously called, so you are not blocked by it.

In the illustration, the execution time moves left to right. I have illustrated the boundary between your main JavaScript thread and the Node.js framework with Libuv (see a later chapter for more information). The upper part is your code being executed in the V8 VM.

## PART II: The Service Layer (Node.js)

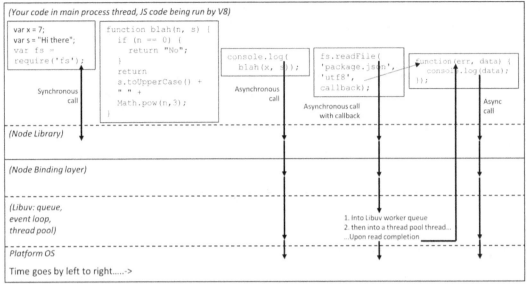

Figure 36 - Code execution flow

Follow left to right in the figure above and you can see how each bit of code is run. To start with, the line of code that does the require() will block code execution until it completes. The console.log() call does not block. The lower layer will asynchronously print out the value. The function blah() is called and in that, the JavaScript Math.pow() object function executes synchronously in V8 and not in the Libuv layer.

The fs.readFile() call uses an asynchronous callback to keep your upper layer code non-blocking. You can see the third parameter to the readFile() function is a function callback. The Node.js framework starts executing readFile() for the first bit of code, but the Node.js code is just calling into Libuv to hand off the request to be executed on its thread pool. This then immediately returns, and your execution continues in the code.

The Libuv execution thread for the file I/O eventually returns and then the callback function gets called and runs on the main thread. You cannot get access to the thread pool processing directly from your JavaScript code.

Filesystem calls go to the thread pool of Libuv. Network calls are processed differently than filesystem calls as will be explained later. In either case, Libuv does all the work for you to make things work across platforms and in a non-blocking way.

# Continuous processing with Node

Another concept that you need to be introduced to is that of how Node can be running in a continuous processing loop. As shown, the previous code sample runs to completion and then

## Chapter 8: Fundamentals

the Node process exits. This is because Node knows if there is any more work to execute and if there isn't any, it just exits.

You can understand that there are certain modules you can use that will basically keep the Node process running forever. If this is the case, you can stop the Node process as you would normally do on your operating system.

You saw code earlier that had a while loop that never had any way to exit. That example would be a little odd, since it runs all of the time, and completely blocks the single processing thread. A more reasonable piece of code would be something that uses an interval timer to do some periodic computation. Here is some simple code that keeps Node running:

```
setInterval(function () {
  console.log('Hello again!');
}, 5000)

console.log('Hello World!');
```

This is actually something similar to what the NewsWatcher code does to periodically look for news stories. NewsWatcher will need to run some processing on a periodic basis.

Another thing that will keep your process running forever would be the use of modules such as HTTP, Net or Express. When you set up the code to listen for TCP connections and listen on a socket, you set up code that will run in the lower level of Libuv. Take the following example that was used before:

```
var http = require('http');

var server = http.createServer(function (request, response) {
  response.writeHead(200, {"Content-Type": "text/plain"});
  response.end("Hello World\n");
});
server.listen(3000);
```

What happens here, is that Libuv is set up to use the low-level OS socket capabilities to listen and respond to incoming connections and requests. Libuv then has a loop to respond to any of these events. When they happen, your callback code can run, such as the one above. The main Node.js process loop actually checks with Libuv to see if it needs to be running because of work it has in its queue, or is listening for. If so, then the process is kept alive.

*Note: the Libuv thread pool is not involved in socket listening. This is because the low-level OS capabilities handle the async non-blocking processing and generates the notification events upon completion.*

# 8.4 Node.js Module Design

Node itself is composed of various modules that run as part of the core service. Modules are what enable all the functionality in Node, besides what is provided with the JavaScript language. The simple "Hello World" example demonstrated this. That example made use of the **console** module. Other modules, such as the **express** module, are third-party modules you install, to bring in additional functionality. As mentioned, you use the Node Package Manager (NPM) to get all your external modules installed on your machine and then use a `require()` statement to use them in code.

You will write many of your own project code as modules for your own consumption. If you are really ambitious, you might even want to write a module and make it available as a download from the NPM repository for others to benefit from.

I will now show you how a module is constructed, and you will see how Node exposes the functionality of a module. You don't necessarily need to know how this works internal to Node.js, but I include this information for those that are curious.

## A module can return a JavaScript object

Modules contain JavaScript code that is typically set up as an object. The object is then exposed in a special way, so that a client can make use of it through the `require()` function as shown in a previous example. Node.js does the work to take your module code and surface it through its internal exposure for other code to call. Node keeps track of all the loaded modules and manages loading, configuring, running, and caching of the modules. All require statements throughout your code will return the same object (i.e. HTTP) as it is cached once and used for every require (singleton pattern).

All you really need to know is that module code is exposed through a special Node object named `exports`. The `exports` object is created for you by Node in each and every module file. Then, when code calls the `require()` function, Node returns the `exports` object with whatever functionality was placed in it, such as a function. The following example shows a simple module you could write that exposes a single function:

```
// mymodule.js that holds your module code
module.exports.welcome = function(name) {
  console.log("Hi " + name);
}
```

As I mentioned, you provide functions and properties in a module file. These are then exposed outside of that file. You can add properties to the exports object such as the `welcome()`

function in the previous example. What Node did for you in the above code was to create the exports object when it ingested your file.

Node has code for the `require()` function that sets everything up to be exposed. Node takes your code from your file, passed in as a parameter, and does something analogous to the following:

```
function require(file) {
  module.exports = {};

  ...The file parameter is used and that file is opened
  ...and code is parsed and taken and placed below.

  // Your extracted code
  module.exports.welcome = function(name) {
    console.log("Hi" + name);
  }
  // End of your extracted code

  return module.exports;
}
```

An empty object is created in the first line of the function. That object then has properties added to it, such as the function you see. Finally, the object is returned so other code can call this function off the object.

For the code to use the function in the module, it needs to use the `require()` function. The `require()` function takes the name of the file that has your module code in it. You reference mymodule.js as follows, and call the function you have exposed.

```
// file server.js that uses the sample module
var w = require("./mymodule.js");
w.welcome("Bob");
```

Node actually has an internal `module` object that it creates for each module exposed. There is a lot more that is going on behind the scenes than that, but you don't need to really know any more of the details. You can certainly learn more of the internals if you are curious by reading through the actual source code, since it is an open-source project.

## A more complicated module

You can hang multiple properties on the exports object, such as objects, strings, numbers, arrays, etc. In one of the previous samples, I showed you the use of the HTTP module. It has a `createServer()` function attached to it. This is a design pattern known as the factory design pattern. You don't use the function directly, as was done with the `welcome()` function in the previous example, but call it to get an object that you can then use.

111

You can also expose a class through a constructor function that clients take and construct themselves, or you can go further and provide a function that does the creation for them like the factory pattern I just mentioned. If you have multiple classes to expose, then you would want to use the factory pattern. If you only have one class, then you can expose that with a constructor function.

If you wand to provide a constructor function in a module, it would look as follows:

```
// mymodule.js
var a = require('http');

function Welcome(nameIn) {
  this.name = nameIn;
}

Welcome.prototype = {
  this.name = null,
  showName: function () {
    console.log("Hi " + name);
  },
  updateName: function (nameIn) {
    this.name = nameIn;
  }
};

module.exports = Welcome;
```

The following example shows how this module can be used:

```
var Welcome = require('./mymodule.js');
var w = new Welcome("John");
w.updateName("blah");
w.showName();
```

The `welcome()` function acts as the constructor in this example. The `prototype` keyword in JavaScript allows you to set properties that exist for all instances and are thus not created again for every instance. Modules are single instance cached anyway. As Node keeps references to each that are used and gives the same instance back every time it is required.

Note how I included the usage of the HTTP module to be used by the sample module above. This shows how modules can require other modules for their own functionality and do so with the standard `require()` function. These included modules are not visible outside of that module code. The code using the module can't actually get access to the HTTP module, unless it also had a `require('http')` statement as well.

## Import instead of require with Node version 10

The JavaScript language itself has releases, and is an implementation based on a standard called ECMAScript. Node has an implementation of JavaScript that is on a path to adopt more ECMAScript syntax as that evolves. In the ES6 release of ECMAScript, a new module system was introduced that uses the 'import' keyword. Version 10 of Node takes a step towards implementing this, so you would no longer need to use 'require' statements if you prefer to use the 'import' keyword instead. People writing HTML browser applications are already familiar with this new standard. Here is what the old way looks like compared with the newer way:

```
// Older CommonJS syntax that existed in Node
const sm = require("./somemodule")

// ESM syntax for getting the default export
import sm from "./somemodule"
```

# 8.5 Useful Node Modules

If you go to https://nodejs.org/api/, you will find the official documentation on the core Node.js modules. Glance through what is there so you can keep it in mind if you need to reference it in the future.

Besides the modules that come with Node, there are plenty of other installable packages from NPM. https://www.npmjs.com/browse/star is a site that lists the most popular ones.

Keep in mind that many of the NPM downloads are for code modules you use inside an application and others are tools that you download and run. Some code modules are also for use as express middleware. Some are for other frameworks such as React or Angular.

Here is a very small list of some code modules you might find useful in Node.js code that you write. A few come with Node itself and the rest you download from NPM.

| Module: | Purpose: |
| --- | --- |
| async | Used to force your code to run in a workflow. Instead of doing things like nesting callbacks, you can use async to set up sequential calls. Also used to run parallel functions with the ability to know when all have finished. |
| child_process | For child process spawning and management. |
| cluster | For setting up a cluster of Node.js processes to distribute the load across. |
| events | A primitive module that many others are built on, to emit or listen to events. |

# PART II: The Service Layer (Node.js)

| | |
|---|---|
| express | Web server functionality for configuring routes, serving up static files and providing template data binding functionality. |
| fs | Standard OS filesystem functionality. |
| helmet | For mitigating different types of HTTP security vulnerabilities. |
| http | This module serves a dual purpose. You can use it to set up a listening service for incoming HTTP requests. You can also use it to make outgoing HTTP requests. |
| joi | For performing validation on HTTP request JSON body properties. |
| lodash | A collection of very useful helper functions (see https://lodash.com/docs/) |
| mongodb | Used to interact with a MongoDB database. |
| net | Standard low-level networking functionality for servers and clients. |
| os | Basic utility functions for accessing OS information. |
| process | The standard type of functionality for working with processes on an operating system. |
| request | Simplifies making HTTP/S client calls. |
| response-time | Displays the response time for HTTP requests. |
| socket.io | A higher-level way to build bi-directional communications between clients and servers. |
| stream | A primitive module that many others are built on to provide readable and writable streams on top of data. |
| url | Utility functions for working with URLs. |
| util | Internal utility functions that Node itself takes advantage of that are exposed for you to use. |
| zlib | For compression and decompression. |

# Chapter 9: Express

The E in the MERN acronym stands for Express. Express is a module commonly used in a Node.js application. It is however not a part of the Node.js installation but can be downloaded from NPM and installed separately and integrated into your application. Express is very popular and consistently one of the most downloaded Node.js modules on NPM. The Express module allows you to implement the functionality of a web server. It provides a way to specify the route handling for incoming requests and simplifies response generation.

One of the things Express does for your project is to remove the need for the HTTP module that comes with Node. The HTTP module that is built into Node.js requires you to write a lot of code in order to set up the routing and responses. Express makes all of that easier.

The following are the key capabilities of Express:
- Specify the route handling of incoming HTTP requests.
- Mechanism to inject middleware into a request to modify it as it gets passed along.
- Easy integration of third-party middleware to provide extended capabilities for request processing.
- Provides a `request` object with properties and methods to look at everything connected with the incoming request.
- Provides a `response` object to use to set everything up for a response.
- Configuration for JSON payload serving.
- Configuration to serve up static files.
- Pairs up with server-side template engines to set context and return HTML with data binding.

## 9.1 The Express Basics

As with any external Node module, you have to do an NPN install and require Express in your code. You can do this at the top of a file such as your server.js file. You can then make the Express listener active. Here is some simple code to set up and use Express:

```
var express = require('express');
var app = express();
app.listen(3000);
```

I will walk you through the construction of the NewsWatcher sample application and show you how to set everything up. Before that happens, I will cover the basics of how Express is used.

PART II: The Service Layer (Node.js)

# Express configuration settings

As part of using the Express module, you will need to configure some settings in your code. These determine how it works and are settings you just need at startup time. You use the Express object `set()` function to do this. The `app.set(name, value)` syntax is used for setting a value for predefined values that Express uses as configuration. The `set()` method will configure values such as what port to listen on. Here is an example:

```
var express = require('express');
var app = express();
app.set('port', 3000);
```

To disable a setting, you can call `app.set(name, false)`. You use the `app.get(name)` method to retrieve any set value. There are `enable` and `disable` functions, but they are the same thing as calling set with a true or false. The following settings can be used for configuring the mode of operation of Express:

**Settings with Value options:**

| case sensitive routing | A Boolean to determine route interpretation as to case sensitivity. For example, if set to false, then "/News" and "/news" are treated the same. |
|---|---|
| env | A string value that sets the environment mode such as "development". This is purely for your use to set and read. For example, you would have logic to determine which URL endpoints, database connections, etc. depending on if you are running the code to try out new development code, or if the code is running in production. |
| etag | Used to set the ETag response header for all responses. Set it to `strong`, `weak`, or `false` if you want to disable it. You can also pass in a custom function. The default value is `weak`. |
| jsonp callback name | A string to specify the default JSONP callback name such as `?callback=`, which is the default. JSONP to bypass the cross-domain policies in web browsers. Not necessarily needed. |
| json replacer | A string to specify the JSON replacer callback. This is a function you define to decide on a property-by-property basis if they are returned on a JSON route response. |
| json spaces | Specifies the number of spaces to use for JSON indenting for readability. |
| port | The port to listen on. |
| query parser | You can set this to `simple` or `extended`. `simple` is based on the query parser from Node and `extended` is implemented by the qs module and is the default. |
| strict routing | A Boolean to set to `true` if you want routes like "/News" and "/News/" treated as different paths. The default setting is `false`. |

## Chapter 9: Express

| subdomain offset | A number that defaults to 2 for how many dot-separated parts to remove to access the subdomain. |
|---|---|
| trust proxy | To be used if you have a front-facing proxy. You will need to set this up to be trusted. This is `false` by default. |
| views | The directory(s) to use for template lookups. |
| view cache | A Boolean that enables template caching. If the `env` setting is set to `production`, then this value defaults to `true`, otherwise, it defaults to `false`. |
| view engine | A string to specify the template engine to use with your provided templates. |
| x-powered-by | A Boolean (defaults is true) to enable the "X-Powered-By: Express" HTTP header to be returned. You would set it to `false` so as not to return it to prevent any hackers from knowing too much about your implementation. |

The Express object has a property named `locals` on which you can place your own custom properties that you might want to associate with your Express application. You use the `app.locals` object as shown in the following example:

```
app.locals.emailForOps = 'Help@myapp.com';
```

## Listening

With code in place to manage your settings, all you need to do next is make a function call to allow the Node.js runtime to run its platform-specific code and set up a socket connection for the given port on the server machine to listen on. Any incoming connections to the IP address of the machine, at that specified port number will be bound through to your code to respond to. Here is the code to do that:

```
var server = app.listen(3000, function() {
   debug('Express server listening on port ' + server.address().port);
});
```

With some understanding of the initialization and settings, you can look at implementing a fully functional web service with HTTP request route handling.

## 9.2 Express Request Routing

When an HTTP request comes into Node, the request will make its way to the Express code you have written to service it. Your Node.js instance will be running on a machine that is hosted and exposed on the Internet. Node will be executing in a process on that server, and through the Express code it would be listening on a port socket for incoming connections.

## PART II: The Service Layer (Node.js)

HTTP requests can come to your server to render a browser page or fulfill REST web service API requests that deal with JSON payloads. A URL request would be like the following:

http://newswatcherscale-env.us-west-2.elasticbeanstalk.com/news?region=USA

If you use query strings as shown above, you can get those values un the Express request handlers. The following diagram illustrates Express routing:

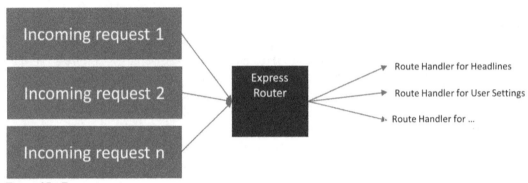

*Figure 37 - Express routing*

If you really need to, you can set up the code to service lower level TCP or UDP types of connections. You can research the Node Modules Net and UDP. This chapter will only be concerned with servicing HTTP requests with Express.

## Routes

Once you have the Express JavaScript object through the require statement, you can use methods that set up the servicing of HTTP requests. Express will then listen for connections on the specified port and it can understand the various HTTP verbs and break down what is being passed in as part of the URL path, query string, and the HTTP headers and body.

The Express object gives you functions to use for handling the various HTTP verbs (get, put etc.). Here are the standard verbs that would be used for CRUD type operations to expose an API in your service layer.

```
app.get(path, callback);    // Read item(s)
app.post(...);              // Create a new item
app.put(...);               // Replace an item
app.patch(...);             // Update an item
app.delete(...);            // Delete an item

function callback(req, res) {
   res.send("Send something back");
};
```

## Chapter 9: Express

Express supports the following HTTP methods:
- `checkout`
- `copy`
- `delete`
- `get`
- `head`
- `lock`
- `merge`
- `mkactivity`
- `mkcol`
- `move`
- `m-search`
- `notify`
- `options`
- `patch`
- `post`
- `purge`
- `put`
- `report`
- `search`
- `subscribe`
- `trace`
- `unlock`
- `unsubscribe`

Please see Express's documentation for the complete list of supported methods. The general signature looks as follows:

```
app.<METHOD>(path, callback [, callback ...])
```

`<METHOD>` would be replaced by one of the methods such as `get`. The first parameter is the URL path. This is not the complete URL, but the portion of it after the domain name.

The second parameter is the callback function that gets called for that request. You can actually provide multiple callbacks and they will each get called sequentially. I will cover more on this later.

Route paths can be strings, string patterns, or regular expressions. They can also be an array that combines any of the mentioned formats. As a string, a route path can be things like "/books". Be aware that the query string portion is not considered part of the path.

## PART II: The Service Layer (Node.js)

You can use "/" to specify that all paths should be picked up. Or, if you omit the path parameter completely, all paths will be picked up.

The callback function you provide has at least two parameters in its standard form. The first parameter is the request object, which contains information about the incoming request. The second parameter is the response object, and is used to send back a response, such as serving up an HTML file or sending back a JSON payload.

The following example shows the use of a pure REST style URL. You just need to provide the routing path that occurs after the domain portion of the URL. Here is the URL and the way you would specify the path.

```
// http://mysite.com/news/categories/sports
app.get("/news/categories/sports", callback);
```

For each verb, such as `get`, you might have different routes for different resources that are being retrieved. Here is some code that sets up multiple path routes for the `get` HTTP verb, each returning something unique:

```
app.get('/about', function(req, res) {
   res.send("About page");
});

app.get('/news', function(req, res) {
   res.send("News page");
});
```

Be aware that the ordering of your route handling code is very important. Any incoming request is basically consumed by the first path that is found to handle it.

## A single path for multiple verbs

If you find you have several verbs that all respond to the same path, you can specify them together by using the `route()` method. This might help you alleviate typing in the path multiple times. An example of this is:

```
app.route('/customer')
   .get(function(req, res) {res.send('Get a customer');})
   .post(function(req, res) {res.send('Add a customer');})
   .put(function(req, res) {res.send('Update a customer');});
```

## All verbs at once

Besides the standard verbs, there are also the methods `all` or `use` to respond to all verbs of the incoming requests. The first route handler below is for all verbs that are for the path

/test. Then the second is set up for everything else and returns a 404 - Not Found code. The use function here is not using a path, so it is for all verbs and all paths not serviced yet.

```
app.all('/test', function(req, res) {
   resp.type('text/plain');
   resp.send('This is a test.');
});

app.use(function(req, res) {
   res.type('text/plain');
   res.status(404);
   res.send('404 - Not found');
});
```

## Advanced path specification

So far, you have seen the simplest of cases with route handling paths. There is a technique that can give you more advanced parsing capability for a URL path. Below is one of the URLs from a previous example:

```
// http://mysite.com/news/categories/sports
app.get("/news/categories/sports", callback);
```

The problem with this example is that you might need to service 20 different categories of news stories. For example, what about the paths /news/categories/science or /news/categories/politics? It could get very monotonous to set up routes for each and every one. To make this easier, Express allows you to set up placeholder parameters in the path that you can then get at later in the callback code and then have the code handle it.

To be able to retrieve parts of a path, you use a special syntax in the path parameter by placing a colon character in the string. This then sets up a JavaScript property you can later access in the handling function as part of the request object. The code below sets the path up with a colon. A property will then be available on the request object as shown.

```
// http://mysite.com/news/categories/sports
app.get("/news/categories/:category', function(req, res) {
   console.log('Your category was ' + req.params.category');
});
```

With the above code, you just have the one route that can service all requests for news stories and can just feed that category into the backend retrieval mechanism.

The req.params property can be used as an array. In JavaScript, that means you can also access it as shown next. The following example shows this, and also shows that you can have more than one of these parameters set up to use:

```
app.get('/products/:category/:id', function(req, res) {
  console.log(req.params[0] + req.params[1]);
});
```

You can also construct your URLs so that they contain query strings. For a query string that you want to process such as `/news?category=sports`, you don't need any special syntax in the path parameter. Just specify the path up to the start of the query string and stop there. The name-value pairs of the query string will automatically be parsed out and made available to you as part of the `query` property of the request object. For this example, there would be a property named `category`, with a value of `sports`:

```
// http://mysite.com/news?category=sports
app.get('/news', function(req, res) {
  console.log(req.query.category);
});
```

## 9.3 Express Middleware

At this point, you know how to set up callback functions that get executed for a given incoming request. Now I'll introduce the concept of middleware as a means of inserting route processing code that will happen before your ending route handler runs. Some middleware code will be provided by Express, other middleware can be by modules downloaded from NPM, and other middleware can from what you write.

The concept of middleware is that you can have code run as an inserted step that is placed before your actual route callback gets run. Middleware code gets chained together in a series of calls that you specify. Express Middleware can then act upon and modify the request object that is being passed along the way. You can do this to reuse code across multiple routes.

Some inserted middleware terminates the request, as it takes care of everything and the requests never even go to any of your route handling. An example of this is the Express static file serving middleware. You might also have some middleware that intercepts calls to verify a user's authorization and does not continue if the request is determined to be invalid.

Another example of middleware might be something that caches content for you. Another example would be a function that logs all operations.

Many available middleware modules are simple to add, yet very powerful in what they provide. I'll show you several of them in this chapter. Conceptually, you can take the Express routing diagram shown previously and modify it as follows to show middleware being injected that passes functionality down the chain:

# Chapter 9: Express

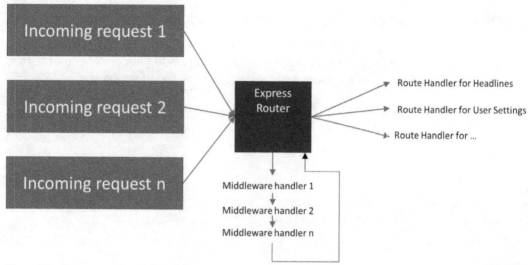

*Figure 38 - Express routing with middleware*

Middleware functions look almost identical to what you have seen already as callbacks. They are just callback functions with the same signature you have seen and thus have access to the request and response objects. This means that the request object can be modified before being passed along. For example, the request body in the response object could be modified to have additional data added to it before it gets to its final destination-handling callback.

You can read the Express documentation to learn about available modules you can download from the NPM repository. Middleware can accomplish things like authentication, caching, logging, session state, cookies, etc. These act as shared pieces of code that you can use across all or just certain routes.

Middleware extends the capabilities of Node.js beyond its core functionality. For example, let's say you are writing a web server that will serve up static files, such as image files. Node.js allows you to do that if you write the code to do so. However, there is a module that acts as middleware in Express that makes it incredibly easy to implement. The module that does this is the `static` module and it comes as part of the Express install. It can be hooked up to serve up static files with very little code.

*Note*: *Don't confuse the concept of a middleware module with the general concept of a module in Node.js. They are still provided through the* `requires` *function, but are used differently than regular Node modules.*

## Hooking up your own custom middleware

Normally, route processing stops at the first match that is found for a URL path and then a callback is run. However, if you make one minor modification to your code, you can string together multiple callbacks for the same route. Notice the one modification made below:

```
// First handler for the route
app.get('/test', function(req, res, next) {
  ...
  console.log("Got here first");
  next();
});

// Second handler for the same route
app.get('/test', function(req, res) {
  ...
  console.log("Got here second");
});
```

The difference is that the first route callback function has a third parameter named `next`. This parameter is a function, and its usage tells Express that you want the callback to act as middleware code to be injected before the actual end route is called. The order is important as stated before.

`next()` is a function that you call when your middleware code is done with all processing. You must call `next()` or the request will be abandoned and not make it to your end handler. In the example, the first handler runs and then, because of the `next()` function call, execution continues to the second handler.

The previous example code could also be structured so that the callbacks are not separated out. You then just list out callbacks one after another. You would need to have the code in the `cb1()` function that calls `next()` or `cb2()` will not be run.

```
app.get('/test', cb1, cb2);

function cb1(req, res, next) {
  console.log("Got here first");
  next();
});

function cb2(req, res) {
  console.log("Got here second")
});
```

## Universal middleware

You can hook up middleware that will get inserted into every single route and for every single verb. To do this, you simply use `app.use()`. This sets up Express to use this function across all incoming requests. You can leave off the path in this case, as "/" is the default path if you don't provide one. Of course, you may want to provide a path so that the middleware only gets run for a certain path.

```
app.use('/', function(req, res next) {
   ...
   next();
});
```

You can insert as much middleware as you need for your routes. The order in which you list them in your code will be the order in which they are sequenced through. Be aware that certain third-party middleware from NPM are required to be placed before others. Refer to the middleware's documentation for more information.

I will now show you a practical example of some custom middleware you might want to implement. Let's say that you have a special path that you only want administrators to have access to. You can create a function that does validation before allowing the request to proceed for further processing. Here is how you do that:

```
// Middleware injection
app.all('/admin/*', doAuthentication);
app.get('/admin/stats', returnStats);
app.get('/admin/approval', approval);
```

With the `app.all()`, the authentication happens for all verbs and acts as middleware. Inside the `doAuthentication()` would be code to determine the authenticity of the request. If it was detected to be invalid, then you would not call next() and the other two route handlers would never be called. You will see something similar being done in the NewsWatcher application.

If you call `next()` and pass an error object parameter, then that route terminates from being handled normally. Error handling middleware then gets invoked. This type of error handling will soon be explained.

## Parameter middleware

Express allows you to set up a middleware callback function for a given parameter property you defined in other route handlers. You do this with the `app.param()` function. This callback is called before any route handler. Here is an example:

```
// Using the global param handler
app.param('id', function(req, res, next, value) {
  console.log("someone queried id " + value);
  if (value != 99)
    next();
});

app.get('/products/:category/:id', function(req, res) {
  console.log(req.params[0] + req.params[1]);
});
```

If you don't have any routes with "id" in them, then the `param()` callback will never get called. The `param` callback will be called before any route handler in which the parameter occurs. You still need to call `next()` to continue the processing.

## Router object

If you really have a lot of routes and want to subdivide them for better organization, you can use the `router` object and keep each in their own modules. In that way, you can isolate your logic and not have it affect the other routes you have set up.

You set the `router` objects up independently. Until you activate them with `app.use()`, they will not be functional. The following is an example of setting up some middleware and some routes in for two different routers and then activating them in the app:

```
var routerA = express.Router();
var routerB = express.Router();

// Set up routes that will end up with news
routerA.get('/weather', function(req, res) {...
});
routerA.get('/sports', function(req, res) {...
});

// Set up routes that will end up for blog
routerB.get('/tech', function(req, res) {...
});
routerB.get('/art', function(req, res) {...
});

// Sets up /news/*
app.use('/news', routerA);

// Sets up /blog/*
app.use('/blog', routerB);
```

## Middleware error handling

Any middleware function can return an error. It does so by calling `next(err)`, where `err` is an Error object. Execution of the middleware and any subsequent routing is ended and processing of the error takes place. To process the error, Express has a default function that it calls that writes the error back to the client. You have the option of providing your own function or chain of functions for handling middleware errors. If you provide a function or chain of functions, then the default one will not be called. In the following example, note how there are four parameters on the error handling middleware function:

```
app.get('/test/:id', function(req, res, next) {
   if (req.params.id == 0)
     next(new Error('Not Found'));

   next();
};

// A middleware error handling function. err is an error object.
app.use(function(err, req, res, next) {
   console.error(err);
   res.status(500).send('Something bad happened');
});
```

Once execution has shifted to the error handling middleware, you can even return back to regular route processing if you do a `next('route')` call. Doing that will jump you to whatever is the next defined route handler.

The diagram of Express routing can be further expanded to add in middleware error handling.

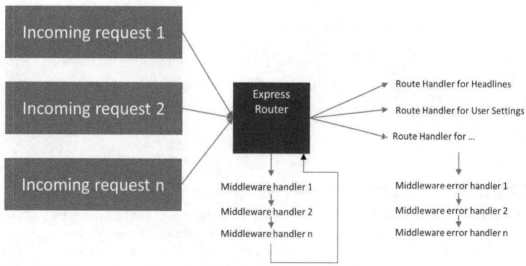

*Figure 39 - Express routing with route handing and middleware error handling*

## Using next() even if your handler is not middleware

I should clear one thing up for you though. Just because you add a `next` parameter on a route handler does not mean you are implementing some middleware. For example, you will most likely need `next()` as a parameter on all your end route handlers in order to do central error processing.

As an example, the following code is an end route being handled by the Express router object. As it is written, you assume nothing will go wrong and just carry out some operation:

```
var router = express.Router();

router.delete('/:id', function (req, res) {
   res.status(200).json({ msg: 'Logged out' });
});
```

What if you wanted to detect an error and use middleware error handling as explained in the previous section? This is where you will need to add the `next` parameter to be called if there is an error. If there is no error, the route will complete by calling the `res.status()` function to send back a response. Nothing else gets in the way, and if running normally, you don't call `next()` because there is nothing else to chain.

In order to use the error handling middleware, an end route itself must be able to pass control to the error handling chain. To do that, the code needs to be modified to add in the `next()` function to be called.

As mentioned, the ending route handler never calls `next()` without giving it an error parameter. The following example shows the additional error handling. The error handler could then send a response with an error code and message.

```
var router = express.Router();

router.delete('/:id', authHelper.checkAuth, function (req, res, next) {
   if (req.params.id != '77')
      return next(new Error('Invalid request'));

   res.status(200).json({ msg: 'Logged out' });
});
```

## Static file serving middleware

One middleware module that comes with Express is the `static` middleware that allows interception of requests for files and returns them. This alleviates the need for you to provide an end route handler of your own. For example, if you want to serve up jpg image files, you can use the following code:

## Chapter 9: Express

```
var express = require('express');
var app = express;

// Middleware injection
app.use('/images', express.static('images'));
app.listen(3000);
```

`app.use` is setting up the middleware route. Instead of the callback function being provided by you, you insert the call for using `express.static()`. This takes care of everything for you to send a response back. In your Node.js project code, you would need to provide the folder of images. An HTML page could access an image as follows:

`<img src="http://yousite.com/images/someimage.jpg"/>`

As a second parameter to `express.static()`, you can pass in an options object that can have the following properties on it: dotfiles, etag, extensions, fallthrough, immutable, index, lastModified, maxAge, redirect, setHeaders.

For example, `setHeaders` is a function you would use to set headers to send with the files. Another example would be setting the `lastModified` property to true and the `Last-Modified` header value is set to the date of the file being sent. For more information, refer to Express's documentation.

## Third-party middleware

There are a growing number of third-party modules that provide Express middleware for your applications. Go to the Express site (http://expressjs.com/resources/middleware.html) to find a list of modules you can download.

To use the middleware, you typically call `app.use(<middleware>)`. This means that all paths and verbs will flow through it. Most third-party middleware that intercepts routes before you get them will call the `next()` function so that processing will eventually reach your code, if you have that need.

Here are a few useful Express middleware components for your reference:

| | |
|---|---|
| passport | Used to authenticate requests. You can set this up to log a person in using OAuth (i.e. through Facebook), or federated login using OpenID. There are more than 300 strategies available through the passport module. |
| body-parser | This middleware intercepts any HTTP Post verb requests that have body data, such as from a form submit. The middleware code runs and |

| | |
|---|---|
| | then by the time your handler code runs, the response object has a body property with sub-properties off of it for each of the body values.<br><br>```<br>var bodyParser = require('body-parser')<br>app.use(bodyParser());<br><br>// In your handler, you can look at the values<br>app.post('/', function(req, res) {<br>  console.log(req.body);<br>}<br>```<br><br>You can specify that JSON is to be parsed and placed in the body. Query string values can also be placed into the body object for you.<br><br>```<br>var bodyParser = require('body-parser')<br>app.use(bodyParser.json());<br>app.use(bodyParser.urlencoded({ extended: true }));<br>``` |
| compression | This will compress requests that pass through the middleware. You would place this statement before any other middleware or routes, unless you only wanted certain routes to be compressed. If you look at the headers of a returned response, you would see the headers have an entry for Content-Encoding.<br><br>```<br>var compression = require('compression')<br>var express = require('express')<br>var app = express()<br>app.use(compression())<br>``` |
| cookie-parser | This middleware does all the work to make available a cookie property on your request object.<br><br>```<br>var cookieParser = require('cookie-parser')<br>app.use(cookieParser());<br><br>// look at all of the properties on body<br>app.post('/', function(req, res) {<br>  console.log(req.cookies);<br>}<br>``` |
| errorhandler | This accomplishes the sending back of stack traces to the client when an error occurs. Make sure to only use this when running in a development environment.<br><br>```<br>var errorhandler = require('errorhandler')<br>app.use(errorhandler());<br>``` |
| express-session | Server-side session data storage. |
| express-simple-cdn | Usage of CDN for static asset serving with multiple host support. |
| helmet | Helpful for mitigating several HTTP security vulnerabilities. |

## Chapter 9: Express

| | |
|---|---|
| response-time | Response time tracking to add the X-Response-Time header. The value inserted is in milliseconds. You could use this to track your SLA over time and be alerted as to what needs further investigation for performance optimization. |
| morgan | This is for request logging. This frees you up from writing any of your own `console.log` statements. For example, incoming HTTP requests go through this middleware and its logs those requests to the console window. You can also specify the format of the logging and direct the output to a file.<br><br>```\nvar fs = require('fs')\nvar morgan = require('morgan')\n\n// create a write stream (in append mode)\nvar accessLogStream = fs.createWriteStream(__dirname + '/access.log', {flags: 'a'})\n\napp.use(morgan('combined', {stream: accessLogStream}))\n``` |
| multer | Multi-part form data, or uploading files in chunks. |
| serve-favicon | For customizing the icon in the browser. |
| timeout | For setting a timeout period for HTTP requests. |
| express-validator | For validation of incoming data. |
| connect-redis | Session store using Redis cache. |
| connect-timeout | For routes that might run into some backend processing issues and need to be limited in the amount of time they take, you can use this to cut them off and return an error. You still need to determine what the right approach is for termination and resubmission of requests.<br><br>```\nvar cto = require('connect-timeout')\n\napp.get('/some_questionable_route', cto('5s'),\n  function(req, res, next) {\n    ...some possibly long running code...\n    ...check req.timeout to see if it is ever true and\n    ...then return false\n    return next();   // finished processing in time, go to next function\n  },\n  function(req, res, next) {\n    res.send('ok');\n  }\n);\n``` |
| vhost | For routing by hostname different sub-domains. i.e. www.mysite.com versus api.mysite.com. |
| express-stormpath | User storage, authentication, authorization, SSO, and data security. Will work with the Okta API. |

# 9.4 Express Request Object

Let's look more in-depth at the usage of the request object in Express route function handlers. The request object contains all the information you need for digesting the incoming request. For example, you have seen how the properties `req.params` and `req.query` are used. You have also seen some third-party middleware that adds more properties to the request object. Here is a reference to some of the properties that are available. You can refer to the Express documentation to find the complete list.

The request object is a parameter of your express route callback function. You can name it anything you like. "req" is a good name for it. The following example shows how to get the complete URL that this request came in from.

```
app.get('/', function(req, res) {
  console.log(req.url);
});
```

Here is a reference to the properties that are available on the `request` object. Refer to the Express documentation for the complete list.

**Request object properties:**

| | |
|---|---|
| `app` | A reference to the instance of the express application object. |
| `params` | This is used to access route parameters. You need to first have set the route specification string and then you can use the `params` property of the request object.<br><br>```app.get('/user/:id', function(req, res) {<br>  res.send('user' + req.params.id);<br>});``` |
| `query` | Used to get the URL querystring. A property will exist on the query object for each.<br><br>```// For URL "/users/search?q=Smith",<br>app.get('/users/search', function(req, res) {<br>  console.log(req.query.q);<br>});``` |
| `body` | Contains properties of key-value pairs of data submitted in the request body. You need to add the body-parser middleware for it to work.<br><br>```var bodyParser = require('body-parser');<br>app.use(bodyParser.json());<br><br>app.post('/', function (req, res) {<br>  console.log(req.body);})``` |

| | |
|---|---|
| route | This is an object that has properties of the current route such as `path`, `keys`, `regexp`, and `params`. |
| cookies | When using the cookie-parser middleware, this property is an object that contains cookies sent by the request. Each cookie that is attached is a property on the cookies object.<br><br>`req.cookies.someName` |
| signedCookies | Exists if the cookies have been signed and protected from tampering.<br><br>`req.signedCookies.someName` |
| ip | The remote IP address of the incoming request. |
| protocol | Such as `http`, `https`, or `trusted` if setup with a trusted proxy. |
| secure | This has a value of `true` if SSL is in effect. |
| headers | The HTTP headers you can access.<br><br>`req.headers['x-auth']` |
| url | The URL of the request. |
| path | The path part of the request, without the query string. |
| route | An object that contains the matched route and lots of other properties such as the method and function that handled it. |
| hostname | The host from the HTTP header. i.e. example.com |
| subdomains | The subdomain part that is in front of the hostname. i.e. ["blah", stuff"] if from stuff.blah.example.com. |
| xhr | This is set to `true` if the request came from a client call such as from XMLHttpRequest, which had set the X-Requested-With field. |

Here is a reference to some of the methods that are available on the `request` object. Refer to Express's documentation for the complete list.

**Request object methods:**

| | |
|---|---|
| `get(field)` | To get the request header fields.<br><br>`req.get('content-type');   // i.e. "text/plain"` |
| `accepts(types)` | To check if a certain type is available, based on the Accept header field. If what you send in as a parameter does not match one of the values, then you will receive an undefined return.<br><br>`req.accepts('html');` |
| `is(type)` | To find out what type the incoming request is.<br><br>`req.is('text/html');   // i.e. returns true` |

| `acceptsLanguages( lang [, ...])` | Based on the Accept-Language field of the header, it returns the first language on a match, or false if none are accepted. Similar calls are `acceptsCharsets` and `acceptsEncodings`.<br><br>```\nvar lang = req.acceptsLanguages('fr', 'es', 'en');\nif (lang) {\n    console.log('The first accepted is: ' + lang);\n} else {\n    console.log('None accepted');\n}\n``` |
|---|---|

# 9.5 Express Response Object

Requests are routed to your callback because of a routing path you have set up. You will eventually return a response back to the requester. The Response object is what you use to do that with. You should at least send back an HTTP status code. You might also return some HTML, text, or better yet, a JSON payload.

Methods on the response object are combined and have a cumulative effect on the returned response. The example code below sets the status of 200 OK for a successful HTTP request:

```
var express = require('express');
var app = express();

app.get('/', function(req, res) {
   res.status(200);
   res.set({'Content-Type': 'text/html'});
   res.send('<html><body>Some body text</body></html>');
});

app.get('/test_json', function(req, res) {
   res.status(200);
   res.set('json spaces', 4);
   res.json({name:'me', age: 37});
});

app.listen(3000);
```

In the first route, the Content-Type is set as "text/html" and then the send() method is used to finish the returned response with some returned HTML. The second route returns some JSON.

The response object has many useful properties and methods. Each usage of the response object is used inside a function callback. You can name it anything you like. "res" is a good name for it.

## Chapter 9: Express

Here is a reference to the properties and methods that are available on the `response` object. Refer to the Express documentation for the complete list.

**Response object properties:**

| | |
|---|---|
| `App` | A reference to the Express application. |
| `headersSent` | A Boolean value that is `true` if HTTP headers have been sent. |
| `locals` | Local variables scoped to the request, might be identical to `app.locals`. A template can use these for its data binding. |

**Response object methods:**

| | |
|---|---|
| `status(code)` | Used to set the status of a return in the case of an error.<br><br>`res.status(404);` |
| `accepts(types)` | For content negotiation on a return.<br><br>`if (req.accepts('text/html') == 'text/html')`<br>`  res.send('<p>Hello</p>');`<br>`} else if (req.accepts('application/json') == 'application/json')`<br>`  res.send({ message: 'Hello' });`<br>`} else {`<br>`  res.status(406).send('Not Acceptable');`<br>`}` |
| `set(field [, value])` | For setting the fields of the response header.<br><br>`res.set({'contentType': 'text/plain', 'ETag': '123'});` |
| `get(field)` | Retrieves what the setting is for a header field.<br><br>`res.get('contentType');` |
| `redirect([code,] URL)` | The path to redirect to instead of the one it came in at. You can provide an optional status code. A 302 "Found" is the default value of the code. You can also redirect relative to the current URL of the service.<br><br>`res.redirect('http://example.com');` |
| `cookie(name, value, [options])` | Sets a cookie. The name is the identifier of the cookie. The value parameter can be a string or object converted to JSON. The options parameter can set up things like `domain`, `expires`, `httpOnly`, `maxAge`, `path`, `secure`, and `signed`.<br><br>`res.cookie('rememberthis', '1', { maxAge: 900000, secure: true });` |
| `json([body])` | Send a JSON body back. |

135

## PART II: The Service Layer (Node.js)

| | |
|---|---|
| `jsonp([body])` | `res.json({ msg: 'Hello' })`<br><br>Sending JSON with JSONP support.<br><br>`res.jsonp({ msg: 'Hello' })` |
| `attachment([path to file])` | You can set an attachment to be returned. If you pass a parameter, it is expected to be a file. The `Content-Disposition` and `Content-Type` are set for you.<br><br>`res.attachment('path/to/logo.png');`<br>`// Content-Disposition: attachment; filename="logo.png"`<br>`// Content-Type: image/png` |
| `sendFile(path [, options] [, fn])` | Transfers a file.<br><br>`res.sendFile('me.png', {maxAge:1, root:'/views/'}, function(err){});` |
| `end([data][,encoding])` | Use this to end the response without any data being returned.<br><br>`res.status(404);`<br>`res.end();` |
| `format(object)` | Use this if you are going to receive requests for content of more than one type. You can line up multiple pieces of code for each Accept HTTP header type.<br><br>`res.format({`<br>  `'text/plain': function(){`<br>    `res.send('Hi');`<br>  `},`<br>  `'text/html': function(){`<br>    `res.send('<p>Hi</p>');`<br>  `},`<br>  `'application/json': function(){`<br>    `res.send({ message: 'Hi' });`<br>  `},`<br>  `'default': function() {`<br>    `res.status(406).send('Not Acceptable');`<br>  `}`<br>`});` |
| `append(field [, value])` | Adds the specified text and value to the header.<br><br>`res.append('Warning', '199 Miscellaneous warning');` |
| `send([body])` | Sending of an HTTP response.<br><br>`res.send({ message: 'Hello' });` |

## Chapter 9: Express

| `sendStatus(code)` | Sets the response code for the return. `res.sendStatus(200);` |
|---|---|
| `render(name, [, data][, callback])` | Template response sending. See the next section of this book for more information. `res.render('user', { name: 'Tobi' }, function(err, html) { ... });` |

# 9.6 Template Response Sending

One of the things that Express enables, is sending server-side HTML that has been generated from templates that have data bound to them. There are several popular template languages that are similar to HTML markup that are supported through Express. I will highlight just one, but you can investigate others.

Using the Express response object, you can formulate a response to send back with a function named `res.render()`. This function takes a file that contains the template as one parameter, referred to as the `view`. As a second parameter, you can provide the data object that binds to the template. The template that you have loaded through Express binds the data and produces the resulting HTML as the output to pass back on the request.

You can pass a third parameter as a callback function to get the rendered string and process any errors that might have occurred. Here is what the code looks like that utilizes a template to send back as a response to a request:

```
app.set('views', path.join(__dirname, 'views'));
app.set('view engine', 'jade');

app.get('/test', function(req, res) {
  res.render('test.jade', {
     title: 'My News Stories ',
     stories: items
  });
});
```

First, you need to have incorporated the NPM jade module into your project. Then you need to make calls to tell Express what directory the template files are in and what template engine you are using. Express will then internally use the Jade module you have included in your project.

137

## PART II: The Service Layer (Node.js)

Notice the two `app.set()` calls used to configure the use of Jade. The `app.get()` sets up the request route with a handler. It is in that handler function you have the call to render the template with the given data to bind to it.

Refer to Jade's documentation to learn about the template syntax. It uses a curly brace syntax to bind to properties. Here is a Jade template that could take the passed in data context and bind those values to elements:

```
// test.jade file content for the template view
doctype html
html
  head
    title my jade template
  body
    h1 Hello #{title}
    div.newstbl
      each story in stories
        div.storyrow
          a(href=story.link)
            img.story-img(src=story.imgUrl)
            h6 #{story.title}
```

Be aware that, if you use template rendering, you are relying on server-side rendering of your HTML. If you prefer to utilize a SPA architecture for a client-side browser application, you would not want to do this. In part three I will describe how to return HTML that uses React component DOM rendering for the client-side SPA application. You can also do server-side rendering (SSR) with React, either to return all your rendered pages, or just a few of them. A React Native application will also be discussed.

# Chapter 10: The MongoDB Module

This chapter is one of the most important ones in part two of this book. That is because one of the main purposes of a service layer is to provide access to the data layer. To do that, you will be utilizing a Node.js module that has been created to interact with MongoDB on the back end. You will be learning how to use the "mongodb" module from the NPM repository.

It is necessary to first include this module in your package.json file so that it is made available in your project. The usual `require()` statement is then used to make functionality available in your code. I will cover that again when you construct the NewsWatcher sample application.

*Note*: *The mongodb NPM module API is quite massive and it would take a large book to document it all. This book cannot make you an expert in all its usage. For example, there are functions to create and delete collections and perform other administrative duties that I have chosen to perform through the Atlas management portal and Compass app. You should certainly make a quick pass through the API to see what other capabilities it has that you might want to take advantage of. Look for the mongodb module on the NPM site and from there find a link to the documentation.*

## The MongoClient object

To begin with, your code needs to establish a connection to a MongoDB server. To do this, you use the `connect()` function of the mongodb `MongoClient` object. After establishing a connection, there are a lot of useful functions for interacting with a MongoDB collection. The function signature for `connect()` looks as follows:

```
connect(urlConnectionString, [options], [callback])
```

The first parameter of the function is the URL of the service endpoint for your MongoDB instance. The second parameter contains options that can be used for settings on the server, such as for a replica set, etc. The last parameter is your callback function, where you will receive the client object that then allows you to be connected to a database and a collection.

If you have an incorrect URL, you will get an error returned. You can find the URL connection string to use by opening the Atlas management portal. Click the **CONNECT** button for the cluster. Click **Connect Your Application**. You will see the connection string listed. You can click the **COPY** button to capture it. URL-encode your username and password if they contain any characters that are not readily used in a URL without converting them, such as '#' or '@'.

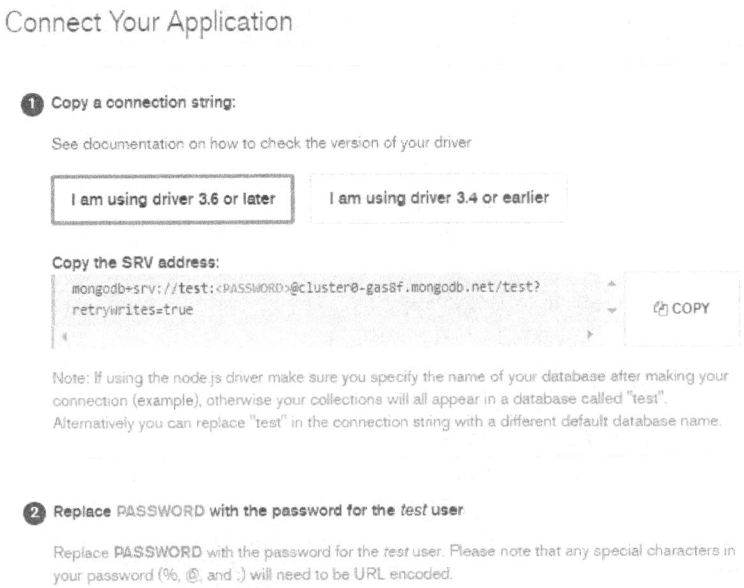

# PART II: The Service Layer (Node.js)

## Connect Your Application

**1** Copy a connection string:

See documentation on how to check the version of your driver

| I am using driver 3.6 or later | I am using driver 3.4 or earlier |

Copy the SRV address:

```
mongodb+srv://test:<PASSWORD>@cluster0-gas8f.mongodb.net/test?
retryWrites=true
```
COPY

Note: If using the node.js driver make sure you specify the name of your database after making your connection (example), otherwise your collections will all appear in a database called "test". Alternatively you can replace "test" in the connection string with a different default database name.

**2** Replace PASSWORD with the password for the *test* user.

Replace PASSWORD with the password for the *test* user. Please note that any special characters in your password (%, @, and :) will need to be URL encoded.

*Figure 40 - Atlas database page with connection url*

In more readable text, this is something like:

```
mongodb+srv://test:pwd@cluster0-gas8f.mongodb.net/test?retryWrites=true
```

There are placeholders there for a password. You should also go to the **Security** tab above the cluster and create a user login that you will use in your code. This is different from the account login you use with the administrative Atlas portal. You need to choose a setting to give the user account read and write access when you create the user.

Here is some example code calling `connect()` to establish a connection using the URL from the above example.

```
var db = {};
var MongoClient = require('mongodb').MongoClient;

MongoClient.connect("<your connection string>", function(err, client) {
  db.collection = client.db('newswatcherdb').collection('newswatcher');
});
```

The code in the callback uses the database connection and calls the `collection()` function to get the "newswatcher" collection object that can then be used to perform CRUD operations.

I can now show you a few of the methods exposed with the mongodb module. The focus will be on learning the functions necessary to perform the CRUD operations.

# Chapter 10: The MongoDB Module

# 10.1 Basic CRUD Operations

With the collection object now obtained, you are ready to learn about the CRUD operations. You can now learn about the four fundamental CRUD operations necessary to utilize a MongoDB database. In the terminology of MongoDB, for single document interactions, CRUD translates to the following functions:

- `insertOne()`
- `findOne()`
- `findOneAndUpdate()`
- `findOneAndDelete()`

I'll now walk you through each of the four functions and show you how to use them. Later, you will put together the NewsWatcher sample application and use them all again, plus a few others. At that time, you will make your code more robust with error handling.

*Note: There are actually variations of the CRUD functions listed above. For example, to find multiple documents, it is necessary to use the `find()` function. When reading any of the documentation, be careful to pay attention to any text stating that it is deprecated.*

## Create

The following is the function signature used for creating a document in a collection:

```
insertOne(doc, [options], [callback] ) -> {Promise}
```

The first parameter is the JavaScript object that you want to have inserted. This then gets created as a BSON document. If you do not provide the `_id` property in your passed in JavaScript object, MongoDB will generate one for you when it stores the document.

Your callback function will have as the first parameter an error object that you can check to see if something went wrong on creation. If you don't provide a callback function, a Promise is returned for you to use. The following code passes an object, and uses the callback function.

```
// This example does not use the options object parameter
db.collection.insertOne({property1: "Hi", property2: 77}, function (err, result) {
   if (err) console.LogError("Create error happened");
   else console.log(JSON.stringify(result.ops[0], null, 4));
});
```

The options parameter of the `insertOne()` function is an object that has a set of properties on it. This same object is used for many of the calls in the API. I will describe it here for your reference, as you will see this used again. All the properties of this object are optional. Here is a table that describes the properties of the `options` object:

**Request Options Object Properties:**

| Property: | Purpose: |
|---|---|
| w | The write concern (how the acknowledgment works). |
| wtimeout | The timeout you want for the write concern. |
| j | To specify the journal write concern. |
| serializeFunctions | Bool to serialize functions on any object. |
| forceServerObjectId | Bool for server assignment of `_id` values instead of driver. |
| bypassDocumentValidation | To bypass schema validation in MongoDB 3.2 or higher. |
| session | Optional session to use for the operation |

The `result` parameter in the callback function has some properties you might be interested in. One of them is the `ops` property. It was used in the previous example and contained the returned document. The `result` properties are as follows:

| Property: | Purpose: |
|---|---|
| insertedCount | The number of documents inserted. |
| ops | An array of all the documents inserted. |
| insertedId | The generated `ObjectId`. |
| connection | The connection object used. |
| result | The command result object returned from MongoDB |

# Read

To retrieve a single document from a collection, use the `findOne()` function. The following is the function signature for retrieving a document:

findOne(query, [options], [callback]) -> {Promise}

The first parameter is the query criteria. You can go back to chapter three to review what that looks like. What this function does is to simply return the very first document that matches the query criteria and no more. The callback has an error object followed by the document returned. Here is an example (not using the options parameter) of `findOne()`:

```
db.collection.findOne({ email: "nb@abc.com"}, function (err, doc) {
    if (err) console.LogError(err);
    else console.log(doc);
});
```

## Chapter 10: The MongoDB Module

Options is an optional object with 20 optional properties you can use. Refer to the mongodb module's documentation for a complete list of properties. The following is a description of a few (there are many more) of them:

| Name | Type | Description |
| --- | --- | --- |
| fields | object | The projection criteria to specify the fields to include or exclude. |
| hint | object | To tell the query what indexes to use. |
| explain | boolean | Return an object with query analysis and not the result. |
| raw | boolean | Return the BSON. |
| readPreference | ReadPreference \| string | Which machine in replica set to read from, such as the primary or secondary. |
| maxTimeMS | Number | How long to wait in milliseconds before aborting the query. |
| session | ClientSession | Optional session to use for the operation |

findOne() returns a JavaScript promise if no callback was provided.

You can use the explain option to look at how well your indexes are working. You will see some JSON returned that gives you useful data about the query. Make sure to remove this option afterward as it prevents your actual result from being returned.

## Update

The following is the function signature for updating documents in MongoDB:

findOneAndUpdate(filter, update, [options], [callback]) -> {Promise}

The first parameter is the filter parameter which is the query criteria needed to identify the document. The second parameter is for the update operators to be used. Your callback function will have as the first parameter, an error object that you can check to see if something went wrong on the update. Here is an example usage that uses an option to have the updated document returned in the callback:

```
db.collection.findOneAndUpdate(
   {email: "nb@abc.com"},
   { $set: { name: "Charles" }},
   { returnOriginal: false },
   function (err, result) {
      if (err) console.log(err);
      else if (result.ok != 1) console.log(result);
      else console.log(result.value);
});
```

## PART II: The Service Layer (Node.js)

The following is a description of the `options` object properties:

| Name | Type | Description |
|---|---|---|
| `projection` | object | The projection criteria to specify the fields to include or exclude in the return. |
| `sort` | object | Species a sorting order for multiple documents that are matched. |
| `maxTimeMS` | Number | How long to wait in milliseconds before aborting the query. |
| `upsert` | boolean | Create the document if it did not exist. |
| `returnOriginal` | boolean | Set this to `false` if you want the updated document returned |
| `Session` | ClientSession | Optional session to use for the operation |

`findOneAndUpdate()` returns a promise if no callback was provided.

## Delete

The following is the function signature for deleting a document in MongoDB:

```
findOneAndDelete(filter, [options], [callback]) -> {Promise}
```

The first parameter is for the query criteria needed to identify the document. Your callback function will have, as the first parameter, an error object that you can check to see if something went wrong on deletion. Here is an example usage:

```
db.collection.findOneAndDelete({email: "nb@abc.com"}, function(err, result)
{
   if (err) console.log(err);
   else if (result.ok != 1) console.log(result);
   else console.log("User Deleted");
});
```

Here is a description of the options object properties:

| Name | Type | Description |
|---|---|---|
| `projection` | object | The projection criteria to specify the fields to include or exclude in the return. |
| `sort` | object | Species a sorting order for multiple documents matched. |
| `maxTimeMS` | Number | How long to wait (milliseconds) before aborting the query. |
| `Session` | ClientSession | Optional session to use for the operation |

`findOneAndDelete()` returns a promise if no callback was provided.

144

## Chapter 10: The MongoDB Module

# 10.2 Aggregation Functionality

In the data layer chapters where the MongoDB capabilities for querying were covered, I omitted one specialized type of query. What I omitted was the capability of MongoDB to perform aggregation over the data. Aggregation gives you the ability to report on summarizations of data such as grouping, or finding the sum, min, or max across a value.

This gets rather involved, so I left it until now to even mention this capability. You really need to make this a focus of some serious study in order to master all that is possible with the aggregation capability.

Here is a simple example to give you a feel for how it works. Imagine that for the example bookstore used in prior examples, you wanted to find out the number of customers living in each state. You would use the `aggregate()` function for this. The following example is the signature of the `aggregate()` function:

```
aggregate(pipeline, [options], callback) -> {null|AggregationCursor}
```

The `pipeline` parameter is an array of MongoDB supported aggregate commands. Think of these as stages the data is being piped through from one to the next. There are quite a few aggregate operators you can string together.

Here is an example that would give the count of people by state. Notice how we chain the function to go to the `toArray()` function.

```
db.collection.aggregate(
  [
    { $group: { "_id": "$address.state", "count": { $sum: 1 } } }
  ]).toArray(function(err, result) {
    console.log(result);
});
```

The result might be as follows:

```
[{ _id: 'UT', count: 54 },
 { _id: 'KS', count: 988 },
 { _id: 'FL', count: 1259 }]
```

The callback has an error object that you can check.

There are operators like `$match` and `$project` that you can insert to help narrow down the documents and what properties are passed through the pipeline.

PART II: The Service Layer (Node.js)

# 10.3 What About an ODM/ORM?

This book shows how to connect directly to MongoDB with a node module created specifically for that purpose. But there is another NPM module you can use that approaches interfacing with MongoDB in a completely different way. That would be with the "mongoose" module.

Those from a relational database background will understand that there are such things as Object Relational Mappings (ORMs) for connecting to a SQL Server. Perhaps you are familiar with Entity Framework for .Net, or Hibernate for Java? The equivalent in a document-based database is called an Object Data Mapping (ODM).

Look up "mongoose" on NPM or GitHub and you will find an ODM that sits on top of MongoDB. This module can be used instead of the mongodb one that we covered in this book.

The mongoose module adds an additional abstraction. With this, you get features like schematization of the data and validation.

Here is what code looks like that uses the mongoose module to save and query a document. I will take the sample of the customer document that might exist in the online bookstore example. I will cut back on the number of properties though, so it is a shorter example. Here is what some code would look like that uses mongoose:

```
var mongoose = require('mongoose');
mongoose.connect('<The usual connection URL>');

var Customer = mongoose.model('Customer', { name: String, age: Number,
email: String });

var c = new Customer({ name: 'Aaron', age: 32, email: 'ab@blah.com' });
c.save(function(err) {
   if (err) console.log(err);
});
mongoose.model('Customer').find(function(err, customers) {
   console.log(customers);
});
```

There are ways to accomplish each of the ODM capabilities on your own. For example, to add in a simple module that helps you do server-side input validation, you can use "express-validator" or "joi". But why go to all the work, if an ODM already exists? An ODM module may be the way to go to give you more robustness with your application.

# 10.4 Concurrency Problems

Once your application starts to have multiple concurrent users, you may run into issues updating documents in MongoDB. Let's take an example where you have a single document that contains a list of high scores for an online game and you want to keep the top five scores across all players. As users finish a game, a call is made to submit the score they achieved and insert it into the list of top scores. High scores are best, so if a new score is posted that should be added because it is higher than other entries, the lowest score is dropped off the list.

If multiple players all submit their scores at the same time, you can understand that some contention might arise with updates to this single document. Here is a diagram of how this might look:

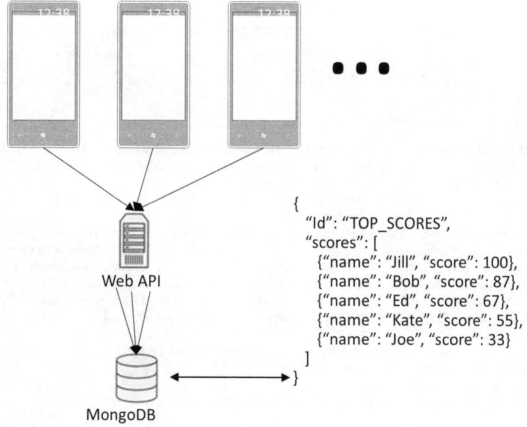

*Figure 41 - Top score contention illustration*

# PART II: The Service Layer (Node.js)

All of these users are causing score insertion code to run in parallel. Each request gets sent to a Node.js Rest API endpoint route to allow multiple simultaneous calls to read and update the single document in the MongoDB DBMS.

What would happen if two scores are submitted for processing at the exact same time? Each request would first read the document and then insert a score if it is higher than the lowest number in the list. The document would then be sent back for replacement in the collection. Here is a sequence diagram that illustrates the problem with time flowing top to bottom:

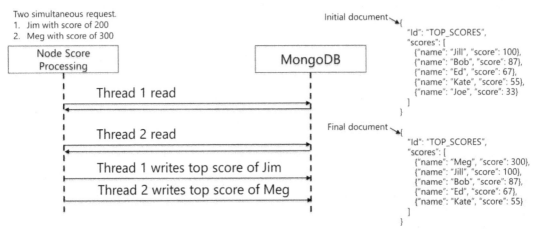

*Figure 42 - Simultaneous update*

If you look carefully, you can see that the final document is not what it should be. You would want to see both Meg and Jim inserted at the top and then Kate and Joe dropped off from the bottom of the array. Instead, Meg is at the top and only Joe dropped off. Jim is really going to be disappointed when he checks the high scores and does not see his name listed.

Neither of the updates knew the other one was happening and so the last write "wins". This is a classic problem and not something unique to MongoDB. There have indeed been many solutions invented over the years to solve the problem. For example, some database technologies offer locking. The problem with this is that locking can be tricky in your own code and requires you to implement detection of stale locks. MongoDB does not support API level locking. It does, however, do this internally on its own for certain calls you make.

The solution that MongoDB and its underlying storage engine (WiredTiger) implement for you is termed optimistic concurrency. If you use the `findOneAndUpdate()` function, this will happen at the document level. If you were, however, to use read, followed by update functions, simultaneous reads will still get the same data and one write will prevail.

With `findOneAndUpdate()`, if two calls come in, the first one pauses the second one from doing anything until the first call has finished on that document. The second call will not even

Chapter 10: The MongoDB Module

do the read until the first one has completed its write of the data. The second call keeps retrying for a time until it is successful.

Let's go back to the initial sequence diagram and draw this out one more time with retries put in:

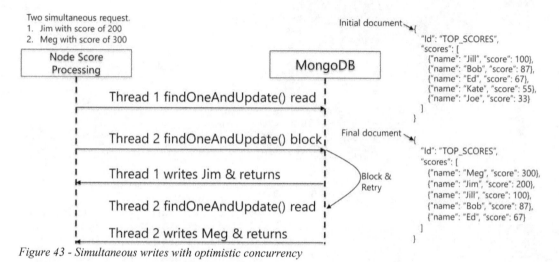

*Figure 43 - Simultaneous writes with optimistic concurrency*

Optimistic concurrency retry logic can be inefficient if your database is always being locked and retries are constantly happening. Imagine if four users consistently update their scores at the same time every second over and over. If four attempts are made at the same time, only one can succeed and then the other three must try again. The second round has three attempts and one works, and so forth.

The point is that you might be wasting MongoDB compute time. This may cause your MongoDB service to perform poorly. You can always look at your design and see if you really need to have one single document that is always being contended for. You may run into cases where you simply have a "hot" document with lots of simultaneous updates all of the time, as was illustrated with the game scores.

Imagine multiple independent Node.js processes that are all hitting a single backend MongoDB database. This means you will have to come up with a way to coordinate across independent processes.

If you want to really implement the ultimate architecture to handle massive scaling and avoid the conflicts you get with optimistic concurrency, you need to create a single queue that is outside of all your node processes.

149

PART II: The Service Layer (Node.js)

You would place your update operation requests in an external queue, such as an AWS SQS resource for global access across all node processes. You then need to create a single unique node process that would watch the AWS queue and process the requests. The following example shows this solution. The dots next to the Scaled Web API represents multiple machines with load balancing across node processes:

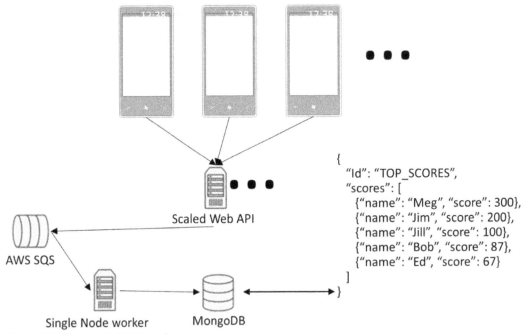

*Figure 44 - Queue serialization solution*

This would also be the safest solution if you really had to ensure consistency and repeatability. There is one main drawback with this solution — you need to deal with the issue of returning to the initial caller if the operation succeeded. This might not be important, but it could be.

You can decide to take the "fire-and-forget" approach and always return success. Perhaps, losing a top score for some reason can be considered acceptable if later the worker role fails to insert a score for some reason.

What if, however, you had a bank transaction that requires a notification one way or the other back to the user? You would need to implement some way to do an asynchronous return. There are notification mechanisms that are available that would solve this. It is possible you could have some callback from the queuing system that could return if the time was in the millisecond range to not delay too much the HTTP request return. You could use WebSockets or HTTP2 capabilities to push back to the browser the result. The request would be queued, and an id returned that the UI can use to poll with to get the completion status.

# Chapter 11: Advanced Node Concepts

There are some particularly difficult issues to be aware of that require careful coordination across the Node ecosystem. There are certain intricacies that must be handled properly. Done incorrectly, it can be disastrous. Done correctly, everything will hum along. This chapter will cover a few of the subtleties you might need to address.

## 11.1 How to Schedule Code to Run

A timer function can be used to schedule code to run at a later time. This can be done as a one-time request, or it can be set up to happen on a recurring interval. You can also look into using the cron module to schedule the running of periodic Node code. The following code schedules a function to run in five seconds and shows how to pass in a parameter:

```
setTimeout(myFcn, 5000, "five");

function myFcn(param1) {
   console.log("Hi %s", param1);
}
```

The callback is not going to happen exactly at 5000 milliseconds, but Node will do its best to fit it in when it is time and the callback is then given to V8 to run.

You can use the function `setInterval()` to schedule a recurring function. If you want to cancel the recurring timer, you can cancel it at any time with `clearInterval()`. The following example shows how this would look:

```
var id = setInterval(myFcn, 5000, "five");
...
clearInterval(id);
```

Be aware that Node will run your function over and over, even if the previous call has been blocked and has not yet completed. If you want a callback executed only if the previous one has completed, you could write your own implementation of `setInterval()` to prevent that behavior.

Another function named `setImmediate()` is available that does not have a timer go off, but places the callback to be executed after I/O, but before `setTimeout()` and `setInterval()` events are processed.

If you are interested in scheduling code to run at an even higher priority, you can use the process object `nextTick()` function. Be careful to not use this unless you are using it responsibly. This call will place your callback to be executed above all other processing in the event queue, even before I/O callbacks are executed. If you do this too often, at some point you will completely cut out all I/O processing.

# 11.2 How to Be RESTful

It is up to you to design the URL endpoints that your Node application will respond to. You can create a RESTful service and even support OData or GraphQL if you like. You are also free to include support for query strings, if that is your preference. If you go the RESTful route, there are a few things to keep in mind.

The first thing to remember is that RESTful web services are intended to be stateless. If you are scaling your server-side Node process and using the cluster module, or have scaling through Elastic Beanstalk, then you most likely want stateless servers. Any of your servers can process any incoming client call. If you really want state information kept around, you can store state information with the client, or have it cached for server-side retrieval in something like a Redis cache.

Make sure to carefully design your REST API URLs so that they make sense. There are standard things to consider such as using nouns and not verbs in your paths. You may run into some dilemmas, but there is probably an answer for your challenge in a forum somewhere.

When you come out with a new version of your REST API, you need to make that apparent and perhaps support multiple versions for a while. You can strive to keep your API as backward compatible as possible. You can insert a version number into your path to move clients from one version to another. For example, if you had "/api/users" as a path, you could have clients start using "/api/v2/users" for new functionality. You can always tack a query string on the end to specify the versioning such as "?Version=2015-12-22". An HTTP header setting is also possible such as "x-version: 2015-12-22".

# 11.3 How to Secure Access

You should conduct a review of all your data connection points and scrutinize all data that is being transferred and stored. Make sure to take appropriate precautions with sensitive data you are safeguarding for your customers. Information such as a home address can be used to identify a person and should never be leaked. Financial and medical records many times have laws and regulations concerning their storage and transmission.

# Chapter 11: Advanced Node Concepts

You are not just trying to ensure your business interests are safe, you are responsible to safeguard your customers from any harm.

One very important detail you need to work out is how users will identify themselves and be allowed to access your Web API from a client application. The other security concern you need to solve is how to prevent any eavesdropping or man-in-the-middle type of hacks as information flows back and forth from the client to the Web API service. This section will explore these and other related topics and present a solution for each.

## Access token

Once a person is identified by their logging in, a Web Service needs to recognize them and allow them access to data associated with their account. One possible means of user interaction authorization is to have them sign in and then use a session cookie with every request coming in.

Another similar mechanism is to generate an access token and have that passed in with every client request. There is a standard way to do this with something called a JSON Web Token (JWT). A JWT is something that can be generated on the server-side in response to a client login request when they present a username and password. The JWT is basically an encoded set of information about the user that is signed to make sure it is not tampered with. Here is the sequence diagram for how a JWT is created and passed from layer to layer:

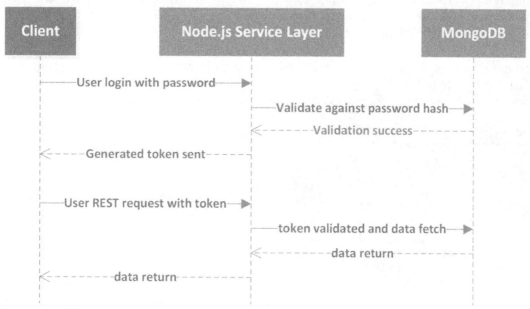

*Figure 45 - JWT token passing*

153

You can pack into the JWT whatever you want. Put in information that would help you in your processing going back and forth. One nice thing would be to add in the IP address as well as the HTTP user-agent header value. This can then be validated on later calls to make sure you still have the same person using the token. You can also set an expiration time on the token so that a person is required to log in every so often.

To offload user authentication, you can use the Oath 2 standard. This allows you to have a person redirected to some other web presence they already trust and log in there first. Users might prefer this, as they only need to have one single sign-on credential to manage.

With the NPM passport module, you can implement Oath 2 to delegate the user authentication to an externally trusted site such as Google or Facebook. A token of authenticity is passed back that has information about who they are.

*Note: The JWT should always be transferred using HTTPS because it can easily be decrypted. Make sure it doesn't contain anything that could compromise your security. The token is in the header such as "x-auth: <token>" or "Authorization: Bearer <token>" if you want to mimic standards like OIDC.*

# All data traffic should be encrypted

One of the first things to do, once you have sufficient momentum on your code, is to implement certificate-based authentication and encryption using the HTTPS standard. This will enable your site to be viewed as legitimate and also ensure that data is transferred using the encrypted TLS/SSL protocol.

If you search on the internet, you will find code samples that show you how to configure Express to require HTTPS. If your Node.js service were to be hosted on a machine directly exposed on the Internet, this is what you would need to do:

```
const https = require('https');
const fs = require('fs');

const options = {
   key: fs.readFileSync('keys/agent-key.pem'),
   cert: fs.readFileSync('keys/agent-cert.pem')
};

https.createServer(options, (req, res) => {
   res.writeHead(200);
   res.end('hello world\n');
}).listen(8000);
```

However, if you choose to use a PaaS solution, the above code is not necessary. This is because your Node.js service is hidden behind the server that acts as the reverse proxy and load balancer. This means that the AWS Elastic Beanstalk service external-facing load balancer needs to be configured for HTTPS. You will see how this is set up when the sample application is put together.

# 11.4 How to Mitigate Attacks

When you expose a service on the Internet, it will be vulnerable to attacks of all kinds. Some attacks might be intentionally malicious and others just annoying. All threats should be taken seriously. At a minimum, they can disrupt your service, which is unacceptable. Beyond that, attacks can steal sensitive information and do damage to your customers and to your own reputation.

Obviously, a simple native mobile phone game of tic-tac-toe would not have as large an attack surface as a three-tier e-commerce application. The more infrastructure and code that you have, the larger your attack surface will be. Hackers will look for the vulnerability that is easiest to exploit. Your security is only as good as your weakest point of attack.

To really get an accurate look at all possible avenues of attack, you need to draw out a data flow diagram that shows all the processes, interactions, data stores, and data flows. Each process would represent those that you own, or ones that you are relying on. Some of those could be classified as completely external and possibly out of your hands, but they should still be on the diagram.

For each of the elements in the diagram, you would want to do some analysis to determine what threats could exist. For example, if you had a SQL database in your diagram, you would determine what it is storing, how data gets in and out and what configuration and administration are happening. You might discover that your SQL database would be vulnerable to a SQL injection attack.

With all of your analysis done, you would mitigate each of the vulnerabilities. In some cases, it might simply involve a few lines of code, in other cases, you might need to re-architect parts of your system. To really address this topic, you should buy a book specifically dedicated to that topic. I will now present a few security concerns associated with Node.js and Web API interactions.

## Never trust ANY input!

One basic strategy to remember is to never trust any input. Always do what you can to validate all data before using it. Look into using the node modules "joi" or "validator". Using a data layer ORM/ODM can give you these data validation capabilities, but you still need to sanitize your data from script injections. Here is some validation using the joi module:

```
var schema = {
  displayName: joi.string().alphanum().min(3).max(50).required(),
  email: joi.string().email().min(7).max(50).required(),
  password: joi.string().regex(/^[a-zA-Z0-9]{3,30}$/)
};

joi.validate(req.body, schema, function (err, value) {
  if (err)
    return next(err);
});
```

You can see that only alphanumeric characters are allowed for the user display name. The joi module is also making sure no extra properties exist on the body. These types of restrictions are extremely important, so don't underestimate their usefulness.

Another issue is data transmission size. What if someone started sending really large JSON packages, or ones that had extra objects or properties in them? You can set up your body-parser middleware to turn down requests that are too large. The default size is 100kb so you can make that smaller just to be safe. Here is how you set that up:

```
app.use(bodyParser.json({ limit: '10kb' }));
```

Let's now looks at some of the types of attacks that could occur on your exposed Web API. None of these are really specific to Node.js. They exist because of the fundamental way that browsers and HTTP work.

# DOS/DDOS attack

The denial-of-service (DoS) or distributed denial-of-service (DDoS) attack is where traffic is thrown at your web app to try to bring it down. It might be possible to overwhelm it so that others are prevented from using it. If successful, the attack will deny service to the actual people that are intended to use it. In some cases, it might actually lead to incorrect behavior of your application, so as to exploit it for other gains.

The distributed version of the attack just means that the attacker is employing multiple distributed machines at once. The term bot is commonly used, meaning that these machines are set up to run scripts or programs that carry out the attack and constantly enlist other machines to also participate. It acts like a virus and replicates.

A DoS attack is purely malicious and would rarely happen just by accident. Regardless, you need to be prepared for it and mitigate this risk. You can be sure that big e-commerce sites like Walmart.com, Amazon, and eBay see these kinds of attacks and take them seriously. An attack like this might even cause your scaling infrastructure to kick in unnecessarily and start costing you more money in cloud operating costs.

# Chapter 11: Advanced Node Concepts

I must, of course, bring up what has already been mentioned – never do anything compute intensive on the main Node.js thread. This is because, if you have a lot of requests coming in that trigger some intensive synchronous code then you will have your process basically unable to respond. This will just make it easier for a real DoS to occur if you let your main thread get overloaded. Let's now look at ways to mitigate a DoS attack.

When you create your Elastic Beanstalk Node.js application environment in the first place, it sets up Nginx to act as the reverse proxy and load balancer. Nginx can be configured to limit the rate per IP address as well as limiting connection count per IP address.

Another approach you can take using AWS is to set up an AWS API Gateway in front of your service layer. That will give you a lot of what you need for defense, such as throttling per connection to head off a DoS attack. Besides that, you also get authorization, reporting, and API consumption of your web API contract in a central shared way.

You could also do some type of IP blocking on your own, such as tracking the access time per IP and then limiting each IP to once per second access. Check out the NPM module express-rate-limit. Of course, you could use a cloud-mitigation provider that would use their expertise to track patterns of attack and identify DDoS attacks and disable them.

## XSS – Cross-Site Scripting

An XSS hack is where some foreign script gets injected and run as part of your web-rendered site. A likely vulnerability would be where you are accepting input from the user and then later re-displaying that input back to them and others that view the site. The browser does not bother to stop JavaScript that was maliciously put in, as it cannot tell the difference. Especially, if you consider that your application allowed the user to enter something in the first place. Normal users will not be typing in malicious scripts to be run, it is the hackers that love to do this for fun and profit.

Let's take an example of something that could affect the NewsWatcher application. In that application, people can comment on a shared news story. This is an occasion where input from the user is accepted and later displayed back to them. Let's say that a malicious user enters the following text as a comment on a news story, instead of some nice comment:

```
<script>alert("Hi");</script><img src="smiley.gif">
```

Now all other users looking at the comment will be affected. In the UI code, you might have some HTML that displays all of the comments and the DOM as follows:

```
<ul><li>
    <p>'<script>alert("Hi");</script><img src="smiley.gif">'</p>
</li></ul>
```

157

Everyone will now see an alert box and also a cute little smiley face staring back at them. You never intended for this to happen, but you did nothing to stop it. The savvy hacker could even hack the client-side JavaScript code and mess with the JSON before it gets sent back to the server. Web APIs simply can never trust the data that is sent to them

Imagine if the hacker referenced some script across the internet that really wreaked havoc. If the hacker understood what your API was on the backend, they could run any command as if they were a logged in person and really do some damage. This means they would have hijacked the user session.

This goes back to the simple statement that you should never trust user input. To mitigate this, you could take action to validate the input as you collect it. For example, in the NewsWatcher code, you validate things like the username and don't allow anything but alphanumeric characters. Your sanitization could also scan all characters and change a character like '<' into "&lt;".

Fortunately, for the NewsWatcher application, when you are using React and render your data, React does the work to disable any scripts from running. React simply displays the actual text without letting the browser interpret it. In the case of the example above, you would actually see the literal string "<script>alert("Hi");</script><img src="smiley.gif">" in the comment list and all would be good.

## CSRF - Cross-Site Request Forgery

A CSRF hack is where a request is made to a site you are currently logged into with your browser. You would have already been authenticated and had an authorization cookie stored by the browser. The attacker would trick you into viewing a page they had set up and as that was loaded in the browser, it would run a script that would send a request to the site you had already been logged onto.

As an example, let's say you were logged on to your banking website. Now, while still logged on, you open up an email that was from a malicious attacker that said: "Click here and win a million dollars!". When you click on the link, the destination URL directs you to their malicious site that loads a page that runs a script that sends requests to your banking site and transfers money to them and changes your password at the same time. Since you are already logged on, the browser happily sends the authentication cookie along with the request and you are hacked. The banking site had no idea that this request was not valid.

Our NewsWatcher sample application does not use cookies, so it is not vulnerable to this attack. The JWT token sending is under the control of your client code and is not automatically sent by the browser like a cookie is.

If you do end up implementing some design that is vulnerable to a CSRF hack, you can use the "csrf" NPM module to implement a mitigation. This will create a secret token that only your site knows that is only sent from your pages.

## The NPM Helmet module

I have discussed each of the security concerns and discussed mitigations. In this section, you will take a look at the Helmet NPM module that would give you the ability to further mitigate possible attacks.

The Helmet module tweaks your HTTP headers to set things up to utilize certain best practices for security risk mitigations. The Helmet module works as Express middleware by injecting itself into the request-response chain. It does not do anything that you can't do by hand. I recommend it though, as it would take you a lot more lines of code for you to accomplish everything that it does with a single line of code.

I will show you some code that will be the starting point when using Helmet. This code will set up things like enforcing HTTPS, mitigating clickjack attacks, certain XSS mitigations, and attacks based on MIME-type overriding attacks.

You can take the defaults and further specify any deviations from there that you like. Refer to the documentation for any of the specific HTTP headers you want to individually control on your own. The following code shows you how easy it is to use helmet:

```
var express = require('express');
var helmet = require('helmet');

var app = express();
app.use(helmet()); // Take the defaults to start with
```

The one usage you do need to control on your own is that which is used with the setting of a Content Security Policy (CSP). This basically lets the browser be aware of where resources can come from. This will then prevent resources unknown to you from being injected. The basic usage of helmet with CSP added becomes the following:

```
var express = require('express');
var helmet = require('helmet');
var csp = require('helmet-csp');

var app = express();
app.use(helmet()); // Take the defaults to start with
app.use(csp({
  // Specify directives for content sources
  directives: {
    defaultSrc: ["'self'"],
    scriptSrc: ["'self'", "'unsafe-inline'", 'ajax.googleapis.com',
```

```
        'maxcdn.bootstrapcdn.com'],
      styleSrc: ["'self'", "'unsafe-inline'", 'maxcdn.bootstrapcdn.com'],
      fontSrc: ["'self'", 'maxcdn.bootstrapcdn.com'],
      imgSrc: ['*']
      // reportUri: '/report-violation',
    }
})); 
```

If you want to have violation notifications sent back to your service, you can uncomment the `reportUri` setting and then handle that Express route in your code. Refer to the documentation for some sample code for that. Helmet cannot be your only mitigation for security threats. You need a thorough analysis of your data flow diagram to come up with every threat and start building your security plan.

## The Node Security Project initiative

There is an ongoing initiative to audit the code of NPM modules and keep a database of known vulnerabilities. This is by no means comprehensive, but it is good that this is underway. Starting with NPM version 6, any install you do will automatically get verified to tell you if there are any known vulnerabilities with the NPM modules you are using. You can also run a command to have an audit run:

```
npm audit
```

Make sure to execute it in the folder where your package.json file is.

# 11.5 Understanding Node Internals

You now know enough to use Node without knowing more about its internal workings. You are thus free to skip this section if you like. You may, however, want to come back to this section later as it might help clear up some advanced questions that might arise.

*Note: This section was written from what I learned by reading through the actual source code for Node.js found on GitHub. Go to https://github.com/nodejs/node if you are interested in the internal workings of Node.js like I was. Libuv code is included in the Node.js source code. You can find out more about its API if you go to http://libuv.org/.*

## The layers of Node.js

You can see from the following block diagram that there exists a main Node.js codebase with dependencies below that. The top two layers of code are what make up the Node.js framework. The bottom two layers are dependencies that Node.js relies on.

# Chapter 11: Advanced Node Concepts

The very top layer is the JavaScript library that you will be using directly from your code. Any time your code does something that is outside of the standard JavaScript calls, you will end up using something that is found in this layer. For example, all of the following code is made possible by the library layer:

```
var http = require('http');

var server = http.createServer(function (request, response) {
  response.writeHead(200, {"Content-Type": "text/plain"});
  response.end("Hello World\n");
});

server.listen(3000);
```

This top layer is what is documented on the Node.js official site https://nodejs.org/en/docs/ as its API. To really get started and use Node, that is the only layer you are really required to know anything about. All core Node modules are exposed through this library layer.

Here are the four layers that comprise the operation of Node as a framework runtime. It is possible that this diagram might change as Node.js changes over time.

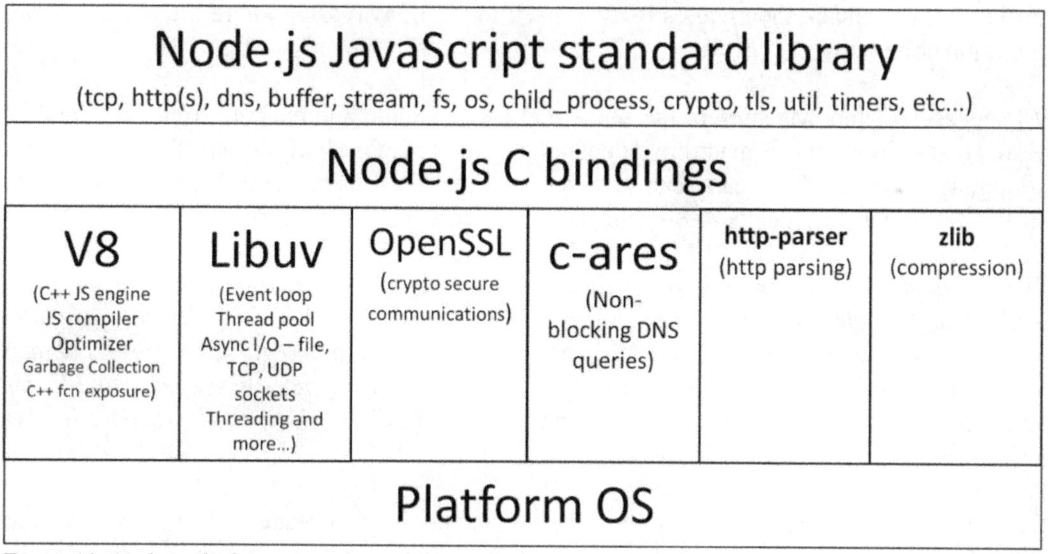

| Node.js JavaScript standard library | | | | | |
|---|---|---|---|---|---|
| (tcp, http(s), dns, buffer, stream, fs, os, child_process, crypto, tls, util, timers, etc...) | | | | | |
| Node.js C bindings | | | | | |
| V8 (C++ JS engine JS compiler Optimizer Garbage Collection C++ fcn exposure) | Libuv (Event loop Thread pool Async I/O – file, TCP, UDP sockets Threading and more...) | OpenSSL (crypto secure communications) | c-ares (Non-blocking DNS queries) | http-parser (http parsing) | zlib (compression) |
| Platform OS | | | | | |

*Figure 46 - Node.js platform*

It would be great if the OS itself could understand JavaScript, but it does not have the libraries exposed for is like it does for C/C++. Thus, there needs to be some translation from your code to code that the operating system can understand. This is why the bindings layer is needed.

161

# PART II: The Service Layer (Node.js)

The Node.js C bindings layer is made up of C++ code. This is what takes your code that is in JavaScript and allows it to call down to code libraries like Libuv that are written in C.

Node uses V8 as a dependency. It does so for two main purposes. V8 makes it possible for your JavaScript code to call through to C++ code running in the Node.js process. Take the following example of some Node.js JavaScript application code you might have. This code displays the size in bytes of your package.json file:

```
var fs = require('fs');

fs.stat("package.json", function(error, stats) {
  console.log(stats.size);
});
```

The way it works is that Node has exposed the fs module in the library layer that is being used here. Node uses some capabilities of V8 to actually take the `fs.stat()` call and have that make a call to a C++ function in the bindings layer called `Stat()`. This C++ function in the binding layer then makes a call to Libuv, which in turn call OS appropriate low-level code. Eventually, it makes its way to a Unix flavor OS library call of `stat()` or on a Windows system, the call made is `NtQueryInformationFile()` that is exposed in ntdll.dll. The callback then makes its way back to your JavaScript where the asynchronous operation completes.

V8 acts as a Virtual Machine in the sense that it can isolate and execute some code and be hosted many times on one machine independently. It provides all of the aspects necessary for a language runtime. You can read an introduction to V8 JavaScript engine by going to https://developers.google.com/v8/intro#about-v8. Here is some text from the google site:

*"V8 is Google's open source, high-performance JavaScript engine. It is written in C++ and is used in Google Chrome, Google's open source browser...V8 compiles and executes JavaScript source code, handles memory allocation for objects, and garbage collects objects it no longer needs...V8 does, however, provide all the data types, operators, objects and functions specified in the ECMA standard. V8 enables any C++ application to expose its own objects and functions to JavaScript code."*

Besides Libuv and V8, there are a few other dependencies that Node uses for operations as noted in the diagram. To actually look at what the dependencies are of Node.js, you can go to the GitHub project and look in the deps folder. Node.js is very portable as its dependencies have been ported to many platforms.

Chapter 11: Advanced Node Concepts

You have now seen how all of the layers fit together. You see that JavaScript code can be executed and that function calls can make their way through Node modules. Some of those modules invoke code through the binding layer in C++ down into Libuv.

The next thing to understand is how the processing loop comes into play. I left out explaining this thus far, but it is important to understand. Understanding the processing loop helps you to be aware of where the processing takes place and how important it is for you to keep your async callback code as performant as possible so as to keep the main thread free.

# Operation of Node

Do not believe everything you read about Node.js. Some people are under the misconception that Node.js only has a single thread and can only do one thing at a time. This is simply not true. I will dispel that misconception and teach you exactly how Node.js executes.

Now that you have seen the layers of code that your application sits on top of, you are ready to see how these layers actually operate to orchestrate the execution flow of a Node application. The main operation of Node is illustrated in the following diagram. Not every dependant component has been included, just the main ones concerning processing flow:

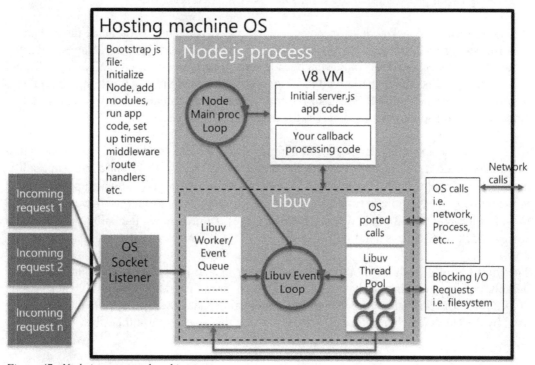

*Figure 47 - Node.js conceptual architecture*

163

# PART II: The Service Layer (Node.js)

It is true that Node.js has a single-threaded architecture for its main JavaScript processing. What is mostly executing on that thread are your JavaScript functions that Node executes in response to events. These are in the form of asynchronous callbacks. This is what gives you concurrent processing in your app. Meaning that more than one operation is executing in the guts of Node and on the operating system in the background, but only one return can and be processed in your code at a time.

Internally, Node can actually use multiple threads to shuttle work off to. Node.js utilizes a callback code pattern which frees up the developer from having to manage the threading and polling themselves. For example, you make your call for retrieving data from a backend database, and as part of that call, you provide a callback function. Your code immediately returns and then Node.js takes care of knowing when the data is returned and schedules your callback to run at a later time. Your callback is thus run asynchronously.

Node.js does not have any sleep, mutex lock, or any similar functions that you might see in other runtimes. On the lower layers of Node, such as in Libuv, the threads of Node can be executing in parallel, taking advantage of running on a multi-core machine. The OS will be able to take the executing threads of Node and spread those out across the cores that are available. Thus, things like file system operations and network requests do happen in parallel, even though your processing of the results can only happen serially.

Your JavaScript callback code is executed asynchronously, meaning you make the call and it is not blocking and code execution can continue elsewhere and you don't know when the return will be processed. It is concurrent because your JavaScirpt code can call many different asynchronous APIs and have them all in action, or queued without you worrying about them. It is parallel because the OS will take the many processing threads, or low-level calls that Node makes and spread them out across the available CPU cores.

If you go back and consider the code sample that used the fs module, you saw that it eventually made it all the way down to the lowest level OS `stat()` function. That call is obviously a blocking call! This is where Libuv does the work for you to queue up that request to take it off your thread and return immediately. Libuv will then run the `stat()` call on a thread in its own thread pool. When the call finishes, Libuv knows the callback to call and sends it back up to be executed in V8.

The thread pool threads of Libuv are being used over and over. There are four of them by default, but you can change that to be more if you find a need. The NewsWatcher sample does not use the filesystem directly, so it does not alter that. The Net module ends up being used by the NewsWatcher database interactions and that does not use the Libuv thread pool.

## Chapter 11: Advanced Node Concepts

# Bootstrapping of Node

If you look back at figure 47 you can see the upper left part has what is termed the bootstrapping of Node. This is something that Node sequences through to get up and running. Here is the general sequence and explanation of the diagram:

1. The Node process is run from a command line and has a `main()` function that is called as its process entry point in its C++ code.
    a. The main function creates the runtime environment and then loads it.
    b. An object called "process" is created that has properties and functions on it. This object is very important and is used throughout the code.
    c. A JavaScript file is now run in the V8 VM that bootstraps the whole process.
        i. Some internal Node modules are parsed and made available.
        ii. The file (i.e. server.js) that node was started with as an argument is read and run in V8.
2. Your server.js file runs and at this point can use the full capabilities of Node to do things like use JavaScript, set up timers, set up HTTP listeners, make HTTP requests, set up middleware etc.
3. After your server.js code is finished being processed, the Node process enters its perpetual processing loop. At this point, the process keeps running as previously stated as long as there is work to process.
    a. Libuv processes work on its queue with its process loop and threads.
    b. Callbacks make it back to the main thread to be run by the V8 VM. V8 would queue calls that come in and process them.

For those of you that are still skeptical about my stating that there are two processing loops, here is the proof. Here is the code from node.cc that the `main()` function eventually enters and keeps executing. This is really what can be called the main processing loop of Node:

```
bool more;
do {
  v8::platform::PumpMessageLoop(default_platform, isolate);

  more = uv_run(env->event_loop(), UV_RUN_ONCE);
  if (more == false) {
    v8::platform::PumpMessageLoop(default_platform, isolate);
    EmitBeforeExit(env);

    // Emit `beforeExit` if the loop became alive either after emitting
    // event, or after running some callbacks.
    more = uv_loop_alive(env->event_loop());

    if (uv_run(env->event_loop(), UV_RUN_NOWAIT) != 0)
      more = true;
  }
} while (more == true);
```

# PART II: The Service Layer (Node.js)

Note how the Libuv processing loop is not allowed to continually run, but is called and is under the control of the main loop of the node process. The main Node process loop calls libuv to let it run with the UV_RUN_NOWAIT flag. The main loop thus is continually calling to have the Libuv loop run over and over. Here is the Unix ported version of the Libuv processing loop:

```c
int uv_run(uv_loop_t* loop, uv_run_mode mode) {
  int timeout;
  int r;
  int ran_pending;

  r = uv__loop_alive(loop);
  if (!r)
    uv__update_time(loop);

  while (r != 0 && loop->stop_flag == 0) {
    uv__update_time(loop);
    uv__run_timers(loop);
    ran_pending = uv__run_pending(loop);
    uv__run_idle(loop);
    uv__run_prepare(loop);

    timeout = 0;
    if ((mode == UV_RUN_ONCE && !ran_pending) || mode == UV_RUN_DEFAULT)
      timeout = uv_backend_timeout(loop);

    uv__io_poll(loop, timeout);
    uv__run_check(loop);
    uv__run_closing_handles(loop);

    if (mode == UV_RUN_ONCE) {
      uv__update_time(loop);
      uv__run_timers(loop);
    }

    r = uv__loop_alive(loop);
    if (mode == UV_RUN_ONCE || mode == UV_RUN_NOWAIT)
      break;
  }

  /* The if statement lets gcc compile it to a conditional store. Avoids
   * dirtying a cache line.
   */
  if (loop->stop_flag != 0)
    loop->stop_flag = 0;

  return r;
}
```

You can see that the while loop has code to cause a break statement to happen in the case of Node calling it.

Callback functions in your JavaScript usage of modules such as with `fs.reafFile()` are kept by Node in a structure. When the low-level Libuv call returns, the callback happens on the C++ main thread of Node.js. If you look at the diagram again, you can see there is a two-way arrow from Libuv to the V8 VM.

The call coming down from JavaScript through the C++ bindings uses the V8 `FunctionTemplate` class to accomplish this. The callback going up from Libuv makes it back to the C++ bindings layer code and uses the V8 `Function::Call()` function to have the V8 VM execute the actual JavaScript callback you had provided.

This callback is not executed on the Node process thread that is orchestrating all of this, but is executed in the V8 VM. Node is in no way managing any queue, event or processing loop for the JavaScript execution. This is all done by V8.

## The effect of compute intensive code

The main event thread should only be used to do fast, less intense processing for inbound or outbound results. Take the following example that shows processing that is unacceptable:

```
var http = require('http');
var bcrypt = require('bcryptjs');
var count = 0;

var server = http.createServer(function (request, response) {
  for (i = 0; i <= 10; i++) {
     bcrypt.hashSync("hjkl5678jhg", 10);
  }

  response.writeHead(200, { "Content-Type": "text/plain" });
  response.end("Hello World\n" + count++);
});

server.listen(3000);
```

Every web request that comes in will cause this compute-intensive code to run. This would basically devour all the compute time as requests keep coming in. Your event loop would not have any time to process any incoming requests.

The point here is that you should never do CPU intensive calculations in your main Node thread callback code. If you do, it will prevent the entire system from working correctly.

PART II: The Service Layer (Node.js)

You can solve this problem in a simple way with Node. There are built-in capabilities to send expensive processing to separate forked processes that do not affect your main Node.js process. Processing can be forked to child processes or even to a separate executable running on your computer that you have written in another language. You could queue up work in a central queue and have Node worker processes pull from that.

You can also create an add-on in C++ for Node.js. You might consider some remote service calls if you have capabilities accessible that way. There are many creative things you can do to solve this problem. As another possibility, you can use AWS Lambda to accomplish the offloading of spot processing or otherwise compute intensive code.

In NewsWatcher we will be forking off code that does the fetching of the global list of news stories. I only do it this way for now to simplify the provisioning and deployment. The code that is running in the forked process, could be taken out and run using AWS Lambda.

# 11.6 How to Scale Node

Could the Node.js process ever get overloaded? For example, let's say 1,000 HTTP Get requests came to your Node.js server all at once. Maybe this, in turn, requires 1,000 files being read from disk. The answer is that Node is set up to be able to manage a sizable workload, so it may do just fine. You would need to do some profiling for your particular workload.

With few exceptions, Node.js makes your requests in a way that is non-blocking. For file system interactions, the requests are shuttled off to threads in the Libuv thread pool to run in parallel. Each of those do their work and return back to the event loop for callback execution scheduling. As thread pool threads get freed up they are given more work.

Eventually, all of the 1,000 file read requests would complete. This happens in the least amount of time possible and also with the least amount of resources. All of this can be handled quite efficiently by one single node process.

You can visualize this queue handoff as shown in the following diagram. Note that the event loop of the Node process simply sequences through everything in its queue if more come in than can be handled. You could configure through code to have more than four threads if you like and you could experiment to see what performance improvements you might have.

# Chapter 11: Advanced Node Concepts

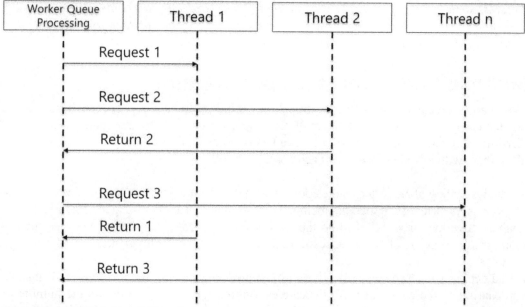

*Figure 48 - Thread pool handoff for file system requests*

Results are always interleaved and processed one by one on your single main thread. As always, the performance is affected by how your JavaScript code processes the results. That is where the bottlenecks almost always happen.

Remember that network I/O is handled differently than file system calls and does not use the Libuv thread pool queue as was discussed for file system usage. Any HTTP calls in your Node code are handled by low level OS mechanisms that can scale.

Be aware that the NewsWatcher sample application will be connecting up to a MongoDB database in the data layer. This module is written to use low-level TCP calls to the MongoDB service. The module you go through to make those calls uses the Net module of Node which eventually calls the library code of Libuv to be making platform network calls. The MongoDB module pools database connections. This means you have a certain amount of connections to use for all the calls in your Node.js code and calls are queued and rotated through as connections are freed up from previous calls. You can configure the number of connections in the pool.

There is obviously a limit to how many operations you can achieve per second before performance starts to degrade. Perhaps you have network I/O requests going off to some slow-returning MongoDB call. The requests could grow out of control if they are not handled fast enough and the DB connection pool requests start to queue up. Of course, you also need to provide for scalability in your data layer, or that will become the bottleneck.

PART II: The Service Layer (Node.js)

*Note: At some point, Node has to rely on the OS and hardware (disks and network cards). Remember, that you have other system contentions to be aware of such as disk controller contentions from other processes running on a given machine.*

## Multiple Node processes per machine

Eventually, you can move to other architectural variations to handle your load. To scale to a greater capacity, you can start up multiple Node.js processes on the same machine and distribute the load across those. If you have lots of cores on a machine, you can make use of all of them with a Node process for each core.

To do this scaling of Node across the cores of a single machine you can use the cluster module of Node.js itself to do the load balancing. There is also a process manager named PM2 that you can download from NPM to do the load balancing. PM2 has other capabilities such as monitoring, restarting of Node processes, and running rolling deployments.

If you deploy your Node.js application through an AWS service, such as Elastic Beanstalk, you can utilize the power of PaaS to scale everything for you by scaling up the number of cores on an EC2 VM (known as vertical scaling) of scale up the number of EC2 instances (known as horizontal scaling). Eventually, you will not be able to handle all requests and processing on a single machine. Ths is when you must start employing horizontal scaling.

## SOA Architecture

The SOA (Service Oriented Architecture) can be understood by contrasting it with what can be considered a monolithic architecture. Perhaps you have seen Java-based or ASP.Net web servers. This is code that renders the HTML on the server and sends it back to the client. It could have a lot of code that is intertwined to spit out this HTML.

This type of monolithic application has many shortcomings. For example, if there was a change to one single part of the code, the whole application needs to be deployed again. Monolithic code is typically tightly coupled and very fragile. The monolith ends up being hard to test, enhance and maintain.

Moving to a SOA, means providing a layer that the UI could access to get at functionality as a set of scrvices. This could be UI that is server-side as well as from a client-side SPA. A great practice that came along was to create these services as HTTP/REST web services that used JSON as the data transfer format.

SOA web services are small units of code that do one thing and are testable. A Node.js application can use Express and have a nice RESTful approach through its exposed paths in the URL that goes to the route handlers. This ends up being a SOA with all of these benefits. You get the benefits of splitting out the capabilities into separate code bases to be separately developed, tested and maintained. Multiple people can work on each individual service independently and each route handler is independent. Each service has their own logic and can go to their own database. They can actually go to the same database, but just have separate document types inside of that shared database.

The code patterns for this type of SOA is what is implemented in the NewsWatcher sample application and can achieve superior quality and scaling. You could deploy the same Node.js application across a horizontally scalable cluster of machines. You could argue that you still have at least one of the characteristics of a monolith. A single codebase that has multiple services still needs a complete deployment for each change that happens in its individual services. The nice thing that makes it different is that each service really is independent and is not affected by a change in another service, so it should not really matter if a new deployment of the collection of routes happens.

## Scaling horizontally

Let's now discuss horizontal scaling. Perhaps you have four cores per machine and have your main Node process on one core. You can count on the OS making use of all cores for you for operations such as file system operations that use the Libuv thread pool.

You could employ the technique of forking processes from the main Node process to offload some heavier computations and to do periodic batch processing. If you are doing this, then you don't have the spare cores to use through cluster or PM2 usage. The better approach is to spread out horizontally across machines. You must do this anyway for really high scale needs.

Scaling across machines is easy to configure in AWS. If you use Elastic Beanstalk, it uses Nginx as a reverse proxy and load balancer for your Node.js application. Elastic Beanstalk lets you configure how many machines to load balance between. You can also set up Elastic Beanstalk to use an auto-scaling group instead of using individual EC2 instances.

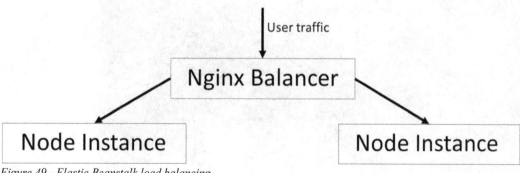

*Figure 49 - Elastic Beanstalk load balancing*

Each EC2 machine would also be in a different availability zone to give you redundant failover if needed. This is in addition to the load balancing benefit that gives you more capacity. Figure 49 shows what that looks like with two EC2 VM instances.

One nice thing with AWS Elastic Beanstalk, is that when machine OS upgrades or patches need to happen, it is done one machine at a time. Machines are taken out of rotation, updated and then put back into rotation. When you need to update your Node.js application, your deployment also happens in this exact same way with a "rolling update". This way you achieve what is called a "zero downtime" deployment.

*Note: Be aware that with any of these scaling mechanisms, you must ensure that you are running in a completely stateless way and avoid any affinity settings to truly be able to distribute calls across all Node processes. To manage a connection requiring state, you can insert the use of something like Redis to fetch state if needed, or keep state only on the client or in a token that is passed back and forth.*

## Static resource serving

In the NewsWatcher sample application, the Node.js application acts to serve up the HTTP requests for the API routes. These are the RESTful web services that the React UI consumes. The Node.js application serves up the actual React web site. These are what are called static resources, because they don't change. All files, such as images, css, html etc. required for the React application are contained in a completely different directory that Node knows about and serves up. For example, here is the logging that happens when the React application is requested. You can see the JavaScript bundled file for React that is being sent. The 304 code you see means that a resource was not send since the version requested was still the latest.

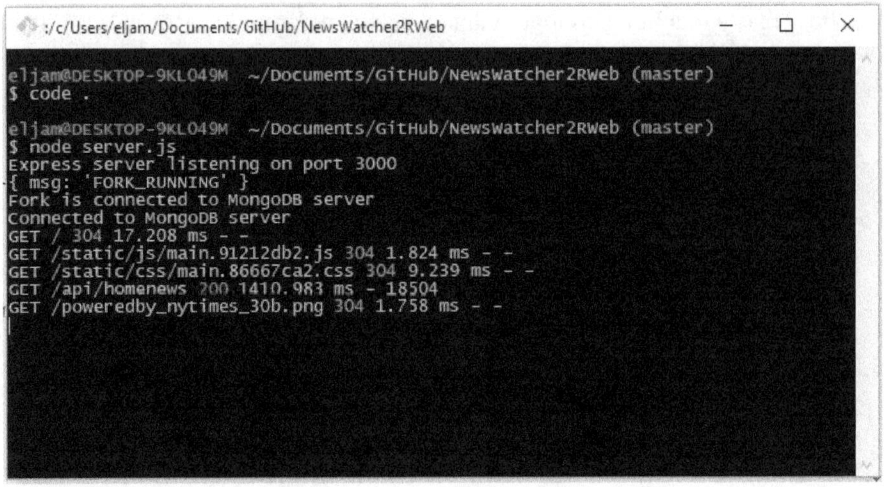

*Figure 50 - Static resource returns*

There are more efficient ways to serve up the React application and free up the Node.js application from that work. For example, the Nginx reverse proxy load balancer can actually know about those requests and serve them up. Another way would be to use something like Amazon S3 storage and set up AWS Route 53 to go through AWS CloudFront that pulls the React application from the S3 storage. This is what is referred to as a CDN (Content Delivery Network).

In the case of Nginx or CloudFront, they both support caching. Since the React application resources are not updated often, requests would be fetched from the cache and be extremely fast. Express.static does not do any caching, but you can use client-side caching with `ETag` or `Max-Age`. By default, the static middleware has `ETags` enabled.

## A Docker Container for your Node.js process

It is possible to use Docker as a means of deploying and running your Node.js application. Elastic Beanstalk can even use a self-contained Docker container. You get the same provisioning, load balancing and scaling you get by just deploying a straight Node.js application.

You might want to take this approach if you find that you have other things to install alongside the Node.js application that can all be contained in the Docker container. If all you have is your Node.js code to deploy using Elastic Beanstalk, you might not have any benefit in using Docker.

It can be convenient for developers to be working in this way so that there is some structure and predictability as to what is needed for a deployment. This would mean that if you have a Docker container all running and tested on your development machine, it is more likely to also function in production, or at least not be missing anything that needs to be installed alongside it.

*Note:* *You can keep your Node.js application Docker images in AWS ECR (EC2 Container Registry).*

## Microservices Architecture

We discussed that implementing SOA was a great idea and that we have achieved that with our RESTful approach. As was mentioned, the NewsWatcher sample application keeps all of the RESTful routes in the same application. Think about what would happen if each individual HTTP/REST route (i.e. auth, user, etc.) were split into their own Node.js application and deployed separately. This gets us into the topic of realizing an architecture that is called a Microservices architecture. The following figure shows the before and after visualization of how our NewsWatcher application could be made into a Microservices architecture.

## PART II: The Service Layer (Node.js)

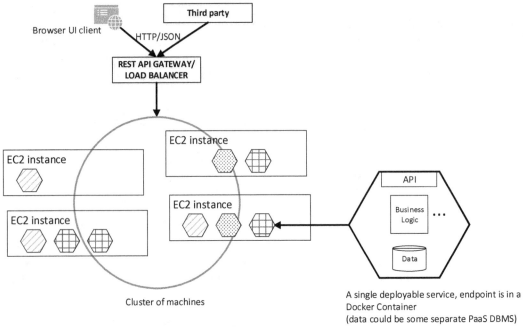

*Figure 51 - Microservices cluster deployment hosting*

Each different background fill pattern in the diagram would represent a different hosted service endpoint. For example, Billing, User Profile, Inventory and such. Each can be independently scaled and the load is balanced across available machines in the cluster. Amazon has ECS for this and also EKS and Fargate.

Microservices are independently deployed and run and each has their own API, middle business logic and backend data layer as necessary. The data could be documents in a shared MongoDB collection. Microservices can be thought of as a specific way to implement SOA.

The term SOA certainly surfaced many years ago and its implementations at that time were very formal and burdened by specifications such as SOAP WS-*, XML/WSDL and were some times implemented on top of an ESB (Enterprise Service Bus). That evolved over time, but the original concepts are still valid in today's modern HTTP REST and JSON.

Before rushing into a Microservices architecture, you should be sure you really need one in the first place. Implementing a full build and deploy system for microservices is more complicated. Always keep things as simple as possible for your needs. Let's look at a few reasons people give for why Microservices are so great and understand under what circumstances you might want to attempt this.

## Chapter 11: Advanced Node Concepts

One benefit is that services are split out into discreet pieces that are much smaller. The argument is that smaller pieces are easier to build and test. You may have heard of companies that manage thousands of microservice at once. Small Microservices provide all of the backend servicing for any number of consumers on the front end. Amazon is one such company that utilizes Microservices. Chances are that you are a much smaller company than Amazon and your applications are much simpler.

Remember to do what is right for your needs. Build your backend services at a granularity that makes development and management easiest for your operation. It is unfortunate that the term Microservices starts with "micro". This makes one think that they must be really small. Indeed, some people advocate microservices that are 100 lines or less. If you really want to do that, then I would advise you go to a serverless architecture and not really do Node.js application, but to look at things like AWS Lambda Functions.

If you were writing a backend billing service, you can imagine that it could be a large amount of code. You can initially try to split it out into smaller consumable units that could each be used independently. If they cannot be used independently, then it should be one cohesive service. At one point, a statement was made in a presentation by Amazon on Microservices that the main amazon.com page makes 100 to 150 backend calls to Microservices at initial load time. These are very small consumable services.

Another benefit that is always brought up is that you can scale better with microservices. That is of course debatable. If you were to take the NewsWatcher application as it is and deploy that on a few dozen machines, you could scale to tens of millions of calls a day. What would happen if you split out the routes into their own deployable Node.js application services and deployed those across the same machines? You might actually get poorer performance because each machine would be running a set of Node.js processes instead of just one. Obviously, the argument is that not all services would be deployed across all machines in the same amounts. Services like billing, might not get called as often and could be deployed on a few machines, while a product lookup service might be called a lot more often and need to be deployed onto lots of machines.

The reason you would want to split out and run the separate services would reduce down to the following observations:

1. You have completely autonomous teams (even geographically separated) that don't know about the other services teams and each is creating and deploying services independently.
2. Each service gets deployed on the number of machines required to handle the load it needs to support.

You would monitor the transactions and loading needs of each service and deploy them out to your cluster as needed. The point of a sophisticated PaaS solution for a Microservices architecture is that it would handle this distribution and shuffling of services to balance them across machines.

## Microservices frameworks

If you have split up your Node.js application into several smaller applications, each needing their own "npm start" command to run, then you will want to consider using Docker for each of these discreet Node.js applications. To do that, you will want to use something like Amazon ECS, EKS, or Fargate. There is also OpenShift Online, Kubernetes, Docker Swarm, Cloud Foundry, or NGINX Unit (Application Server and Service Mesh). These frameworks manage the deployment of your Docker containers to a cluster of machines.

Really sophisticated Microservices architectures may even employ some type of service discovery registry. There are frameworks such as Scenica that are available to do this. You need to justify that you really need this complexity and overhead.

There is a Node.js package named Hydra that you can download from NPM that might be of benefit to you is you are creating a more complex implementation. Hydra gives you things like service discovery, distributed messaging, message load balancing, logging, presence, and health monitoring.

## Queues and worker processes

A common approach in backend services is to take requests that require longer running processing and accept that request immediately and return success, but actually just queue it up. Imagine you had a photo storage service that has an HTTP Post endpoint API that accepts images. You might want to create thumbnail images and do some image processing to do facial recognition and add meta data to identify people.

To keep your service scalable, you could have your web service accept the image, copy it to S3 and then place an entry into MongoDB or AWS SQS to be looked at later. Then you could have separate EC2 machines or AWS Lambda functions that scale and independently read the work requests and do the processing. This is how to manage the work and keep the web service very responsive.

# Chapter 12: NewsWatcher App Development

It is now time to begin constructing a Node.js Service layer that integrates with the MongoDB data layer from part one of this book. This chapter takes the concepts you have already learned regarding Node and applies them in a real project. What you will be creating is a RESTful web API that your presentation layer will be able to integrate with. You will learn how to implement everything needed for a fully functional cloud-hosted web service. You will utilize best practices for testing and DevOps in the chapters that follow.

*Note*: Don't forget that you can access all of the code for the NewsWatcher sample project at https://github.com/eljamaki01/NewsWatcher2RWeb.

## 12.1 Install the Necessary Tools

One of the amazing things about Node is how simple and quick it is to set up a server. It can be hosted in AWS as a PaaS offering using Elastic Beanstalk, or other hosting infrastructure, or even be deployed and managed manually. Node.js runs on many different operating systems.

To get started, install the following:
- ✓ A code editor such as Visual Studio Code, Sublime text, Vim, etc.
- ✓ Node.js from https://nodejs.org/. Get a stable version. This installs the node executable for you and also installs the NPM executable.

*Note*: Visual Studio Code is not the same as Visual Studio. VS Code is a completely new tool that offers a rich editing environment as well as integrated features for source code control and debugging. VS Code has the capability to launch tasks through the means of tools like Gulp with no need to jump out to a command line. These tools can be used to automate build and test steps that you need to run frequently. With Visual Studio Code, you will be able to create a project and run it locally on your machine and have access to IntelliSense, debugging, and web app publishing through Git/GitHub.

## 12.2 Create an Express Application

Start by creating a folder for your application. I named mine "NewsWatcher2RWeb". You can now create the minimum amount of code required for a Node.js application. To help maintain your sanity, you should start with the smallest amount of code possible and push it

PART II: The Service Layer (Node.js)

all the way to deployment. This will eliminate unneeded investigation time of issues unrelated to just getting the basic service up and working.

Launch Visual Studio Code and click **File->Open Folder**, then select the "NewsWatcher2RWeb" folder you just created. With VS Code open, you see the **EXPLORE** view open and in there find two subfolders. You can also use a command prompt and type "code ." to open up VS Code in the project directory.

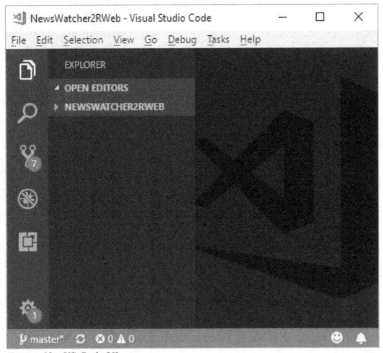

*Figure 52 - VS Code UI*

The **NewsWatcher2RWeb** folder shows you all files and subfolders.

You can now get started and create a simple Node.js application, and then deploy it to an AWS Elastic Beanstalk EC2 instance. Once that is verified and running ok, you can add more code to fill out the full functionality of the web service.

On the NEWSWATCHER subfolder, click on the icon to create new files. Start by creating these three files:

- .gitignore
- package.json
- server.js

## Chapter 12: NewsWatcher App Development

The .gitignore file will not be used right now, but would come into use when you make use of Git and GitHub. You list files and directories you want to exclude from Git source code control.

At this point, there is no node_modules folder yet. It will automatically be created when you install the node modules with the "npm install" command.

Add the following code to the server.js file:

```
var express = require('express');
var app = express();

app.get('/', function (req, res) {
  console.log('Send message on get request');
  res.send('Hello full-stack development!');
});

app.set('port', process.env.PORT || 3000);

var server = app.listen(app.get('port'), function () {
  console.log('Express server listening on port:' +
              server.address().port);
});
```

Add the following lines to the **package.json** file:

```
{
  "name": "NewsWatcher",
  "version": "0.0.0",
  "description": "NewsWatcher",
  "main": "server.js",
  "author": {
    "name": "yourname",
    "email": ""
  },
  "scripts": {
    "start": "node server.js"
  },
  "dependencies": {
    "express": "^4.13.4"
  }
}
```

Save all the files, then open a command prompt window and navigate to your project's directory. At the command prompt, type `npm install`. This will look at your package.json file, and install the standard Node modules along with the Express module that is listed as a dependency. A new directory will be added with the name `node_modules`.

179

PART II: The Service Layer (Node.js)

You can now try running your Node project locally. Once it is proven to function, you will work on getting it deployed to AWS. At the command prompt, type "`npm start`". You can also type "`node server.js`" to run it. If the project runs successfully, you will see the following console output:

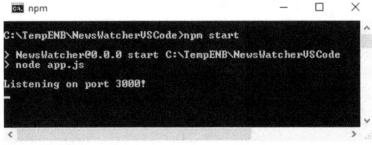

*Figure 53 - local console output*

Open a web browser and navigate to http://localhost:3000/. In the browser window, you will see your message:

*Figure 54 - Project message in a browser window*

# 12.3 Deploying to AWS Elastic Beanstalk

It is now time to create your Elastic Beanstalk app through the AWS Management Console. To do so, you must already have an AWS account. ***Be aware that you may incur some cost at this point if you are not running with a free account with AWS!***

To create the app:
1. Open a web browser and navigate to https://console.aws.amazon.com/console/.
2. In the upper right corner of the web page, for **Region**, select a region such as **US East (N. Virginia)** as the region you want your services to be running in.

Chapter 12: NewsWatcher App Development

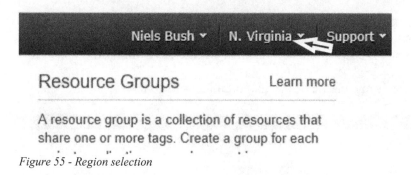

Figure 55 - Region selection

3. From the selection of services, click **Elastic Beanstalk**:

Figure 56 - Select Elastic Beanstalk

4. Click **Create New Application**. Enter a name and description. I gave it the name "newswatcher".

Figure 57 - Elastic Beanstalk create new application

PART II: The Service Layer (Node.js)

5. Click to create a new environment by clicking **Create one now**. Select a **Web server environment**.
6. Select **Node.js** as the platform and click **Create environment**.

Once the site is ready, the elastic beanstalk portal looks as follows:

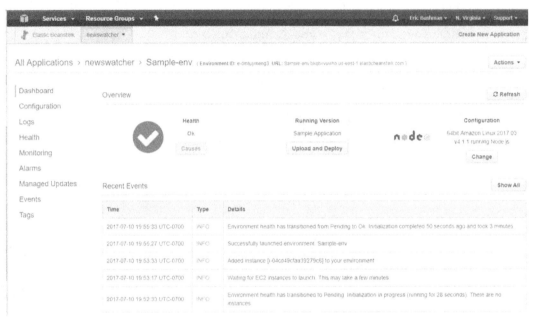

*Figure 58 - Elastic Beanstalk portal*

Click the **URL:** link at the top of the page to see your site working.

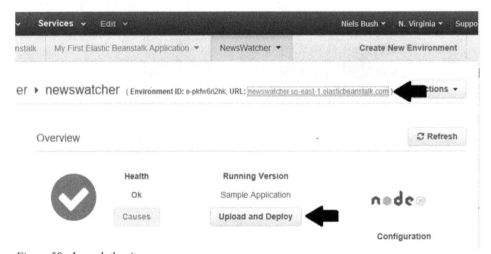

*Figure 59 - Launch the site*

## Chapter 12: NewsWatcher App Development

Now you can deploy your own simple Express application. On a windows machine, this is as follows:
1. Using the Windows File Explorer, navigate to your project folder.
2. Select the package.json and server.js files together, then right-click and select **Send to -> Compressed (zipped) folder**. Give the zip file a name and save it.
3. In the Elastic Beanstalk dashboard for the newswatcher application, click **Upload and Deploy** button and select your zip file.
4. Wait for the confirmation that the deployment is ready and click the URL again.

Your Node.js application is now working for everyone to see.

You do not need to zip and send the node_modules folder. The deployment to Elastic Beanstalk will run 'npm install' for you on the EC2 instances. One thing you should do is to set an environment variable through the Elastic Beanstalk management console so that the node install actually becomes "npm install –production". Set the environment variable in the **Configuration -> Software Configuration -> Environment Properties**. as:

```
Property name:NPM_CONFIG_PRODUCTION   Property value:true
```

This will make the install go much faster as it will not deploy any npm modules that are needed in test or development environments.

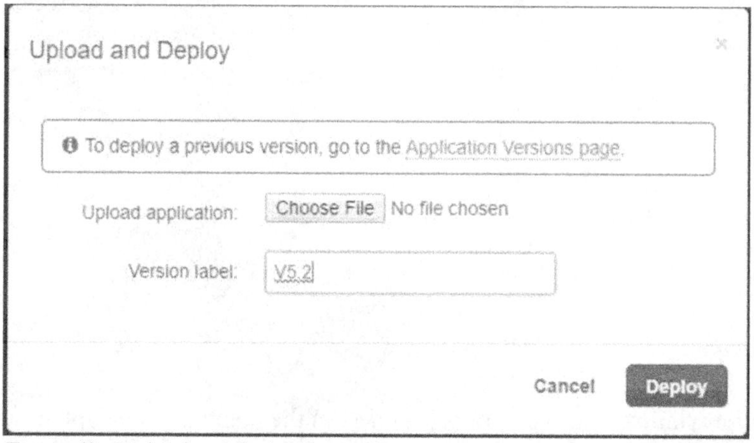

*Figure 60 - Upload and deploy the zip file*

You just performed a manual deployment. While a few manual deployments might be tolerable, you eventually want full continuous-integration scripts that run tests and deployments for you.

PART II: The Service Layer (Node.js)

# 12.4 Basic Project Structure

You can now add the rest of the code for the NewsWatcher application. You can start by adding in the rest of the Node.js dependencies that you will need. Edit your package.json file to be as follows, then save it. It should look similar to the following. You can refer to the GitHub project for the complete file.

```
{
  "name": "newswatcher",
  "version": "0.0.1",
  "main": "server.js",
  "scripts": {
    "start": "node server.js",
  },
  "dependencies": {
    "async": "^2.6.0",
    "bcryptjs": "^2.4.3",
    "body-parser": "^1.18.2",
    "dotenv": "^4.0.0",
    "express": "^4.16.3",
    "express-rate-limit": "^2.11.0",
    "helmet": "^3.12.0",
    "helmet-csp": "^2.7.0",
    "joi": "^13.3.0",
    "jwt-simple": "^0.5.1",
    "mongodb": "^3.0.8",
    "morgan": "^1.9.0",
    "response-time": "^2.3.2"
  },
  "devDependencies": {
    "eslint": "^4.19.1",
    "eslint-plugin-react": "^7.8.1",
    "mocha": "^4.1.0",
    "selenium-webdriver": "^3.6.0",
    "supertest": "^3.0.0"
  }
}
```

Now open a command prompt window and type `npm install` at the command prompt. You will add code to the rest of the files later in this chapter. You can, of course, go to the GitHub project and get all of the code for the NewsWatcher sample application. You are now ready to put your service layer web application RESTful API together.

***Note:*** *If you ever want to completely update to the latest version of one or more installed packages in your package.json, change every version number to "\*", and then run – "npm update -–save" and "npm update –save-dev".*

# Chapter 12: NewsWatcher App Development

# 12.5 Where it All Starts (server.js)

The starting point of your Node application can be in a file named whatever you like, I prefer server.js. This file instructs Node what to do to be initialized and what to subsequently execute, once up and running. In reality, this file could be named anything. This file does things like establish a MongoDB connection, set up error handling, and establish the Express listener for the HTTP request handling.

At the top of the server.js file is where you will place the `require` statements that load the needed modules. If you recall, module references inside a file are internal to that file.

Some of the modules you will specify as being required are actually used as middleware for Express. This means you set them up with a `require` statement and then don't actually use them directly. They act to intercept calls through an `app.use()` setting. If needed, you can refer back to the chapter on middleware to review this concept.

Here is the first section of code in the server.js file. I have added comments to briefly describe the purpose of each require statement.

```
var express = require('express'); // Route handlers and templates usage
var path = require('path'); // Populating the path property of the request
var logger = require('morgan'); // HTTP request logging
var bodyParser = require('body-parser'); // Access to the HTTP request body
var cp = require('child_process'); // Forking a separate Node.js processes
var responseTime = require('response-time'); // Performance logging
var assert = require('assert'); // assert testing of values
var helmet = require('helmet'); // Security measures
var RateLimit = require('express-rate-limit'); // IP based rate limiter
var csp = require('helmet-csp');
```

The first require statement provides the object that will be needed for leveraging Express. The path module is basically used to provide a helper object that will be used to manipulate strings for specifying the file paths in your project. The rest of the require statements are for setting up modules that act as middleware.

These next lines to add will set up environment settings for the code that can have things like passwords and other secrets and configurations that are needed. When run locally, this is read out of a file named .env. When run in production, the file is not used, but the values are set in the AWS Elastic Beanstalk environment settings and the code sees those instead.

```
if (process.env.NODE_ENV !== 'production') { // reading in of secrets
  require('dotenv').config();
}
```

# PART II: The Service Layer (Node.js)

These next lines will pull in code from modules that you will write yourself that provide the route handlers:

```
var users = require('./routes/users');
var session = require('./routes/session');
var sharedNews = require('./routes/sharedNews');
var homeNews = require('./routes/homeNews');
```

The code can now start to implement some of its capabilities. You now make the Express application object available and set a setting on it that will be needed for when it is run in AWS. Since the app is behind an Nginx load balancer with Elastic Beanstalk, you don't want the load balancer IP address sent in the header requests, but want the IP address of the actual machine that it was acting on behalf of. This is what the trust proxy setting does.

```
var app = express();
app.enable('trust proxy');
```

## Middleware

Next, you can see how the Express middleware is hooked up for some of the modules you are incorporating. Here are those lines:

```
// Apply limits to all requests
var limiter = new RateLimit({
  windowMs: 15 * 60 * 1000, // 15 minutes
  max: 100, // limit each IP to 100 requests per windowMs
  delayMs: 0 // disable delaying - full speed until the max limit
});
app.use(limiter);

app.use(helmet()); // Take the defaults to start with
app.use(csp({
  // Specify directives for content sources
  directives: {
    defaultSrc: ["'self'"],
    scriptSrc: ["'self'", "'unsafe-inline'", 'ajax.googleapis.com',
      'maxcdn.bootstrapcdn.com'],
    styleSrc: ["'self'", "'unsafe-inline'", 'maxcdn.bootstrapcdn.com'],
    fontSrc: ["'self'", 'maxcdn.bootstrapcdn.com'],
    imgSrc: ['*']
  }
}));

// Adds an X-Response-Time header to responses to measure response times
app.use(responseTime());

// logs all HTTP requests. The "dev" option gives it a specific styling
app.use(logger('dev'));

// Sets up the response object in routes to contain a body property with an
```

## Chapter 12: NewsWatcher App Development

```
// object of what is parsed from a JSON body request payload
// There is no need for allowing a huge body, it might be some attack,
// so use the limit option
app.use(bodyParser.json({ limit: '100kb' }));

// Main HTML page to be returned is in the build directory
app.get('/', function (req, res) {
  res.sendFile(path.join(__dirname, 'build', 'index.html'));
});

// Serving up of static content such as HTML for the React
// SPA, images, CSS files, and JavaScript files
app.use(express.static(path.join(__dirname, 'build')));
```

This code takes the modules brought in through the `require()` statements and inserts them as middleware by calling `app.use()`.

The first piece of middleware is what gives protection against DoS attacks. You can look up the module in GitHub to actually see how it works as middleware.

The next piece of middleware is Helmet. I covered its use earlier. Helmet is a security mitigation module that tweaks the HTTP headers.

There are five other uses of middleware that are documented in the code to tell you what they do. Each is very useful and you will benefit from them all.

You can see the use of the path module to provide functionality to manipulate path strings with the join function. The `__dirname` variable is provided by Node so that you can use it to get the name of the directory that the currently executing code resides in. In this usage, it would return the directory of the server.js file. It will be the local path if you are running it locally, or whatever it is on the AWS production machine if it is running in the deployed environment.

The coding of the React Web SPA React is covered in the third part of this book. You can see the code that serves up this static content that is using the Express static module.

### Forking a Process

The next code in the server.js file is used to fork off a separate Node.js process and give it a file to execute. This is used to shuttle off any code processing that is more intensive and that you don't want run on your main Node process thread.

```
var node2 = cp.fork('./worker/app_FORK.js');
```

Earlier, I explained that you need to be careful with code you write that wills execute on the main Node.js V8 VM. With NewsWatcher you need to offload a few things to a separate

## PART II: The Service Layer (Node.js)

Node process. These involve code for collecting news stories from internet sources into NewsWatcher's master list in MongoDB and doing the filtered matching of stories for users. I will show you the code in the forked process later. Remember also, that I said it would probably be better to use AWS Lambda for this type of code processing.

You can see that you use the child_process module to start up the second process. Since Node is ported to many platforms, it will call whatever low-level code is needed to accomplish this for the OS it is running on. Node makes use of some libraries that do this and these have been ported already. You can pass messages back and forth between processes if you like. You will be doing this later in code.

If the forked process is experiencing runtime errors, it could shut itself down and then the main process could be signaled to start it up again. You add code in server.js to restart the forked process as follows:

```
node2.on('exit', function (code) {
  node2 = undefined;
  node2 = cp.fork('./worker/app_FORK.js');
});
```

**The MongoDB Data Layer Connection**

Most of the code in the service layer deals with interactions with the backend data storage layer. You initialize your MongoDB connection by utilizing the mongodb module that is an NPM download. You use the `connect()` function and then set up the usage of the newswatcher collection. The `connect()` function takes the MongoDB connection URL.

```
var db = {};
var MongoClient = require('mongodb').MongoClient;

//Use connect method to connect to the Server
MongoClient.connect(process.env.MONGODB_CONNECT_URL, function (err, client)
{
  assert.equal(null, err);
  db.client = client;
  db.collection = client.db('newswatcherdb').collection('newswatcher');
});
```

You save the connection as a property on an object you set up named **db** to be used later with your Express routes through middleware injection. Watch for that code coming up soon.

The last thing to note about the above code is that you have a configuration file (.env) for keeping settings in that you want to have in a central place. Some of the values in that file are ones you want to keep secret. Don't post that file for anyone to see. As was mentioned already, you keep needed configuration values there, but these values are also set as environment name/value pairs in an Elastic Beanstalk environment.

# Chapter 12: NewsWatcher App Development

**Sharing Objects**

The `node2` and `db` variables are needed in your routing code. The `db` object database connection will be used for all of the CRUD operations, so you need to make that available. You expose these variables through middleware injection. This means that you have a chance to inject the objects into the request processing chain by adding them as properties on the request object.

You place a middleware function right at the top of the Express chain that every request will have to pass through first. In there you attach new properties on the request object that is being passed along.

As required, you call `next()` to move the execution along to the rest of the processing chain for the request. Remember that the `use` function applies across all requests, so that a `get`, `put` or any other request is routed through here first.

```
app.use(function (req, res, next) {
  req.db = db;
  req.node2 = node2;
  next();
});
```

**Express Route Handlers**

You are almost done with the main application code that sets everything up. The following are all of the routes for your HTTP/Rest API. This goes back to understanding what your objects are and what verbs each will support. You just list out each object. Inside each of the supporting modules you will find the verbs and any sub-objects off of those. These are modules you write that use the Express Router object as explained in section 9.2.

```
// Rest API routes
app.use('/api/users', users);
app.use('/api/sessions', session);
app.use('/api/sharednews', sharedNews);
app.use('/api/homenews', homeNews);
```

Next, there is an error handling route that handles invalid URLs that come in. This returns a 404 code to signal that the resource was not found. Basically, if none of the routes match, then this one will. For example, if a request came in for /api/blah, it would go here. This code activates an express error handler, because it calls `next(err)`.

```
// catch everything else and forward to error handler as a 404 to return
app.use(function (req, res, next) {
  var err = new Error('Not Found');
  err.status = 404;
  next(err);
});
```

PART II: The Service Layer (Node.js)

Here are the error handling routes for when you have an error returned in the code. You get here when `next(err)` is called in any routing code. There is also a handler that only kicks in when running in your development environment. You want to do this so that you can add in what the stack trace is. The second handler is the one that kicks in, in the production environment.

```
// development error handler that will add in a stacktrace
if (app.get('env') === 'development') {
  app.use(function (err, req, res, next) {
    res.status(err.status || 500).json({ message: err.toString(),
              error: err });
    console.log(err);
  });
}

// production error handler with no stacktraces exposed to users
app.use(function (err, req, res, next) {
  res.status(err.status || 500).json({message: err.toString(), error: {}});
  console.log(err);
});
```

The final lines in server.js contain the necessary code that tells Express to be listening for HTTP requests. In production, it picks up the port necessary to run in that hosted environment. The last three lines shown, export the server for our testing framework to use.

```
app.set('port', process.env.PORT || 3000);

var server = app.listen(app.get('port'), function () {
  console.log('Express server listening on port ' + server.address().port);
});

server.db = db;
server.node2 = node2;
module.exports = server;
```

If you look at the code in the GitHub project, you will see that I also added route handlers for starting and stopping the V8 profiler and for taking memory snapshots. These can then be activated by placing a call to that particular route, and also deactivated when not needed.

# 12.6 A MongoDB Document to Hold News Stories

You need to create a MongoDB document to store the contents of the shared global news stories. This is a one-time creation done in advance that can be done through the MongoDB Inc. Compass application. This document functions as the holder for the master list of current news stories and top stories found on the home page. This is then used by all users to do matching of their news filters with. This way, each user does not need to fetch all the news individually. This document will have a distinctive value set for the `_id` property so you know what it is for. Be careful to not accidentally delete it. This document looks as follows:

```
{
  "_id": "MASTER_STORIES_DO_NOT_DELETE",
  "newsStories": [],
  "homeNewsStories": []
}
```

I covered this previously in part one of this book, so you might have already created this document. If not, you can create it at this time. This document could be created in code if you really wanted to. Since it is a one-time thing, I have chosen to use Compass to create it.

# 12.7 A Central Place for Configuration (.env)

To run your Node application locally, there will need to be certain environment variables set that your code can reference. These are values such as those to establish the connection to the MongoDB database. You place an .env file in the project and then use an npm module to consume that. This technique was used in the server.js file as already explained.

Be aware that I cannot divulge the actual contents of my .env file, as you would then have access to services I need to protect. Replace values as needed in your own .env file.

```
// 3.6 mongodb driver and later connection string
MONGODB_CONNECT_URL=mongodb+srv://babajee:mypass@cluster0-k5ghj.mongodb.net/test
JWT_SECRET=<yoursecretkey>
NEWYORKTIMES_API_KEY=<yoursecretkey>
GLOBAL_STORIES_ID=MASTER_STORIES_DO_NOT_DELETE
MAX_SHARED_STORIES=30
MAX_COMMENTS=30
MAX_FILTERS=5
MAX_FILTER_STORIES=15
```

The value for the MongoDB connection URL can be found in the Atlas portal as was explained previously. You go to the clusters page and click the **CONNECT** button and find the string for connecting to an application. Follow the instructions that are found there. It will tell you to replace the placeholders for your own password.

# 12.8 HTTP/Rest Web Service API

It is now time to fill in the REST Web Service API. The API will accept and pass back JSON payloads in response to HTTP requests. Eventually, you will create the SPA web page that calls the web service being designed here.

If you think about the REST API that you want to expose, it becomes clear that you need to create all the CRUD operations for each resource that is necessary. Your resources are sessions, users, sharednews and homenews.

Here is a table that lists everything the REST API supports. The `id` parameter is the identifier of individual resources for a user being accessed. The `sid` parameter is the identifier for stories.

| Verb and Path | Result |
| --- | --- |
| POST /api/sessions | Create a login session token. |
| DELETE /api/sessions/:id | Delete a login session token. |
| POST /api/users | Create a user with the passed in JSON of the HTTP body. |
| DELETE /api/users/:id | Delete a single specified user. |
| GET /api/users/:id | Return the JSON of a single specified user. |
| PUT /api/users/:id | Replace a user with the passed-in JSON of the HTTP body. |
| POST /api/users/:id/savedstories | Save a story for user, content of which is in the JSON body |
| DELETE /api/users/:id/savedstories/:sid | Delete a story that the user had previously saved. |
| POST /api/sharednews | Share a news story as contained in the JSON body |
| GET /api/sharednews | Get all of the shared news stories |
| DELETE /api/sharednews/:sid | Delete a news story that had been shared. |
| POST /api/sharednews/:sid/comments | Add a comment to a specified shared news story |
| GET /api/homenews | Get all the homepage news stories |

# Chapter 12: NewsWatcher App Development

You may have noticed that some verbs are missing that you might have expected to find. For example, you will not see a `GET /api/users` to get the list of all users. You don't really want other people to see everyone that is a user of NewsWatcher, so don't offer that. You certainly could decide to offer it, but you would then want to place a middleware restriction on it that only allows logged in administrators to have access to it.

An example of a restricted API route that you have is `DELETE /api/users/:id`. A user can only delete themselves, so you restrict that to just the logged-in user for deleting their own account and not an account of anyone else. You could write some code to allow admins to be able to delete anyone.

Remember that the token is useful for restricting access with. It is up to you to define what roles and access you will need and then enforce it. In this case, each call only works for that account to access that person's own data for whatever is authorized. Perhaps administrators that log in can be identified and allowed access to everything.

## Visualizing the code

If you recall, the lines below are found in the server.js file and are used to set up your Express route handling. The first two Express application calls are used for sending the files back that the client browser application will need. The last four Express application calls are for route handling of everything listed in the REST API resource table.

Here are the lines from the server.js file as a reminder:

```
// Main file to serve up that is built by React build process
app.get('/', function (req, res) {
  res.sendFile(path.join(__dirname, 'build', 'index.html'));
});

// Serving up of static content such as HTML for the React SPA, images,
// CSS files, and JavaScript files
app.use(express.static(path.join(__dirname, 'build')));

// Rest API routes
app.use('/api/users', users);
app.use('/api/sessions', session);
app.use('/api/sharednews', sharedNews);
app.use('/api/homenews', homeNews);
```

The following is a pictorial representation of how the routing code all hooks together. This does not contain all of the files and details, but it at least gives you an idea of the routes that are being serviced. I have even included the UI rendering React code even though that has not been covered yet. You can see that I divide up the code by the architectural layer it resides in.

PART II: The Service Layer (Node.js)

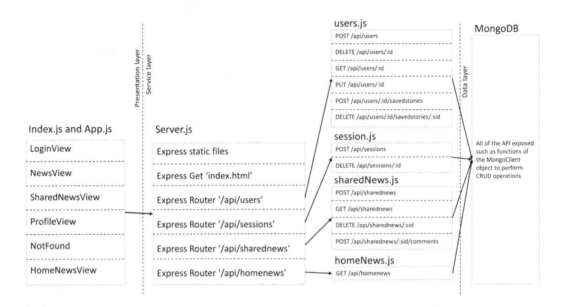

*Figure 61 - NewsWatcher file diagram*

You can now look at the individual files used for each of the Express routes one at a time. They each exist in their own file. These are the files where you find the servicing of the individual HTTP verbs (GET, POST, PUT, DELETE).

# 12.9 Session Resource Routing (routes/session.js)

The session route is used in the API to allow people to log in and out. Here are the specific routes for the session resource:

| Verb and Path | Result |
| --- | --- |
| POST /api/sessions | Create a login session token. |
| DELETE /api/sessions/:id | Delete a login session token. |

The post operation takes a user's email and password in the request body and basically logs a user into their account. A token is sent back in the response body. The client calling code can take the JWT and then keep passing it back on subsequent calls to identify that person. The token can be stored in client-side storage and used as needed.

194

## Chapter 12: NewsWatcher App Development

The `post` verb handler needs to first query for the user document to see if they actually have a registered account. The user must exist, or they cannot be logged in. If there is a match for the email, then the stored password hash is validated with a hash of the password coming in.

If the password is validated, then a token is created and passed back in the JSON payload. The tokens could be used forever, but it would be easy to add a timestamp to set it to expire. With that added, you might do something like require a new login every six months.

As was discussed, there is a bit of verification that can go on with the token. You check it to make sure it is originating from where the token was originally assigned from. The IP address and the header setting for `user-agent` are kept with the token as additional verification.

There is no database storage of the token in the NewsWatcher app, so the route delete does not really need to do much. There is just a simple check to verify that the person logging out is the same as the one contained in the token.

The code uses the joi module to validate the incoming request body object parameters. Here is the code for session.js:

```
// session.js: A Node.js Module for session login and logout
"use strict";
var express = require('express');
var bcrypt = require('bcryptjs'); // For password hash comparing
var jwt = require('jwt-simple'); // For token authentication
var joi = require('joi'); // For data validation
var authHelper = require('./authHelper');

var router = express.Router();

//
// Create a security token as the user logs in that can be passed to the
// client and used on subsequent calls.
// The user email and password are sent in the body of the request.
//
router.post('/', function postSession(req, res, next) {
  // Password must be 7 to 15 characters in length and
  // contain at least one numeric digit and a special character
  var schema = {
    email: joi.string().email().min(7).max(50).required(),
    password: joi.string().regex(/^(?=.*[0-9])(?=.*[!@#$%^&*])[a-zA-Z0-9!@#$%^&*]{7,15}$/).required()
  };

  joi.validate(req.body, schema, function (err) {
    if (err)
      return next(new Error('Invalid field: password 7 to 15 (one number, one special character)'));
```

195

```
    req.db.collection.findOne({ type: 'USER_TYPE', email: req.body.email },
    function (err, user) {
      if (err)
        return next(err);

      if (!user)
        return next(new Error('User was not found.'));

      bcrypt.compare(req.body.password, user.passwordHash,
      function comparePassword(err, match) {
        if (match) {
          try {
            var token = jwt.encode({ authorized: true, sessionIP: req.ip,
                                     sessionUA: req.headers['user-agent'],
                                     userId: user._id.toHexString(),
                                     displayName: user.displayName },
                                     process.env.JWT_SECRET);
            res.status(201).json({ displayName: user.displayName,
                                   userId: user._id.toHexString(),
                                   token: token, msg: 'Authorized' });
          } catch (err) {
            return next(err);
          }
        } else {
          return next(new Error('Wrong password'));
        }
      });
    });
  });

//
// Delete the token as a user logs out
//
router.delete('/:id', authHelper.checkAuth, function (req, res, next) {
  // Verify the passed in id is the same as that in the auth token
  if (req.params.id != req.auth.userId)
    return next(new Error('Invalid request for logout'));

  res.status(200).json({ msg: 'Logged out' });
});

module.exports = router;
```

Notice the use of the middleware function `authHelper.checkAuth`. This is something you will define next. This is what allows you to inject an authorization check before proceeding to the final function that does the work. If the authorization fails then the function for the end route handling will not be called.

# 12.10 Authorization Token Module (routes/authHelper.js)

You have seen how a user logs in and a token is generated. You now need to create some middleware that will be inserted and run for verifying the token before proceeding on a route that needs to be secure. This code will look at the passed-in token and make sure it is valid. As you know, middleware can be inserted into any route and that is what will be happening here.

Each of the routing modules will make use of the `authHelper` module to verify that a valid token is being passed before performing any other action. This happens because each HTTP/Rest call would have an `x-auth` header token value filled in. You could use a header such as `Authorization: Bearer <token>` if you want to mimic standards like OIDC.

You will see the use of the `checkAuth` function in many of the route handlers. The code here will simply verify that there is an `x-auth` header and, if there is, then decode it with the secret, then set the decoded object in a request property named `auth` for further usage by anything in the processing chain. If the token is missing, has been tampered with, or does not contain what it is supposed to contain, an error is returned.

When you look at the password handling code of session.js, you see where all the information was placed in the token. It really is up to you to decide what to put in there. With this token, you can represent the user being signed in. Do not store the user password (or other sensitive data) in the token. Here is the middleware that verifies that a token is valid:

```
//
// authHelper.js: A Node.js Module to inject middleware that validates the
// request header User token.
//

"use strict";
var jwt = require('jwt-simple');

//
// Check for a token in the custom header setting and verify that it is
// signed and has not been tampered with.
// If no header token is present, maybe the user
// The JWT Simple package will throw exceptions
//
module.exports.checkAuth = function (req, res, next) {
  if (req.headers['x-auth']) {
    try {
      req.auth = jwt.decode(req.headers['x-auth'], process.env.JWT_SECRET);
      if (req.auth && req.auth.authorized && req.auth.userId) {
```

## PART II: The Service Layer (Node.js)

```
      return next();
    } else {
      return next(new Error('User is not logged in.'));
    }
  } catch (err) {
    return next(err);
  }
} else {
  return next(new Error('User is not logged in.'));
}
};
```

If you provide a UI check-box for the user to stay logged in, then they can check that you can store the token on the device and not have to log them in each time. If a user wanted to get their token from the returned login request, they could. If they passed it on to anyone else, their account could possibly be compromised. Thus, the extra tests. There is a test for the IP address being the same. This may not actually work if you have users logging in from different devices all the time with different IP addresses.

# 12.11 User Resource Routing (routes/users.js)

The user resource represents information for a logged in user. A user document retrieved by their ID will contain information like their profile that has their news filters and also the news stories that have matched. For a given user, you can also make Rest calls to save a story or delete a saved story. Here are the specific routes for the user resource:

| Verb and Path | Result |
| --- | --- |
| POST /api/users | Create a user with the passed-in JSON of the HTTP body. |
| DELETE /api/users/:id | Delete a single specified user. |
| GET /api/users/:id | Return the JSON of a single specified user. |
| PUT /api/users/:id | Replace a user with the passed-in JSON of the HTTP body. |
| POST /api/users/:id/savedstories | Save a story for user, content of which is in the JSON body |
| DELETE /api/users/:id/savedstories/:sid | Delete a story that the user had previously saved. |

Start by looking at the require statements at the very top of users.js. Some of the modules required are ones that you have seen before.

# Chapter 12: NewsWatcher App Development

```
var express = require('express');
var bcrypt = require('bcryptjs');
var async = require('async');
var joi = require('joi'); // For data validation
var authHelper = require('./authHelper');
var ObjectId = require('mongodb').ObjectID;

var router = express.Router();

...code cut out here...

module.exports = router;
```

Let's look at the verb handling functions one by one.

**POST /api/users**
The `post` verb takes a JSON payload and creates a new user account as a document in your collection. This happens when a user account is first created. A password is passed in as part of the JSON body and you create a hash of it to store in the MongoDB database document for that user.

This call will fail if there already exists a document in the collection with that email value already. Relying on an email address to identify a user account is one way to keep user accounts unique and identifiable. The mongodb `findOne()` function is what is used to see if a user account existed already.

The code goes ahead and creates all of the properties ever needed in the document. Default values are used that make sense. There is even a sample news filter set up for the user.

Notice the call to `node2.send()`. This is used to pass a message to the forked Node process to offload the filter processing for that newly added user. This call will take the new account and do the initial matching of stories for the filter. I have not shown you the code for the forked process yet, but you can see that the concept is very simple.

What happens right at the start, is the validation of the passed-in JSON body. You want to make sure that there are no extra properties, and validate that the allowed ones conform to some known types and have safe values. The joi NPM module is used to perform the validations. Here is the code for the post verb handler:

```
router.post('/', function postUser(req, res, next) {
  // Password must be 7 to 15 characters in length and contain at least one
  // numeric digit and a special character
  var schema = {
    displayName: joi.string().alphanum().min(3).max(50).required(),
    email: joi.string().email().min(7).max(50).required(),
```

199

## PART II: The Service Layer (Node.js)

```
      password: joi.string().regex(/^(?=.*[0-9])(?=.*[!@#$%^&*])[a-zA-Z0-9!@#$%^&*]{7,15}$/).required()
    };

    joi.validate(req.body, schema, function (err, value) {
      if (err)
        return next(new Error('Invalid field: display name 3 to 50 alphanumeric, valid email, password 7 to 15 (one number, one special character)'));

      req.db.collection.findOne({ type: 'USER_TYPE', email: req.body.email }, function (err, doc) {
        if (err)
          return next(err);

        if (doc)
          return next(new Error('Email account already registered'));

        var xferUser = {
          type: 'USER_TYPE',
          displayName: req.body.displayName,
          email: req.body.email,
          passwordHash: null,
          date: Date.now(),
          completed: false,
          settings: {
            requireWIFI: true,
            enableAlerts: false
          },
          newsFilters: [{
            name: 'Technology Companies',
            keyWords: ['Apple', 'Microsoft', 'IBM', 'Amazon', 'Google', 'Intel'],
            enableAlert: false,
            alertFrequency: 0,
            enableAutoDelete: false,
            deleteTime: 0,
            timeOfLastScan: 0,
            newsStories: []
          }],
          savedStories: []
        };

        bcrypt.hash(req.body.password, 10, function getHash(err, hash) {
          if (err)
            return next(err);

          xferUser.passwordHash = hash;
          req.db.collection.insertOne(xferUser, function createUser(err, result) {
            if (err)
              return next(err);
```

## Chapter 12: NewsWatcher App Development

```
                req.node2.send({ msg: 'REFRESH_STORIES', doc: result.ops[0] });
                res.status(201).json(result.ops[0]);
            });
        });
    });
  });
});
```

Users are allowed to sign up for an account by having them provide their username, email, and a password. You will not store the actual password, but will instead store an encrypted hashed value of the password. Even if anyone were to get a hold of that for a user, they would still not be able to log in with it as it is extremely difficult to decrypt that into the password.

Once a person is registered as a user, then the `/api/session/` path will be used to accept their email and password to get their session going each time they want to log in and use NewsWatcher.

As explained in the chapter on authentication and authorization, you will be sending a token in the header of each HTTP/Rest request. When a user logs in, they get a token that is subsequently used to identify them for all further interactions.

**DELETE /api/users/:id**
With this path, you see the use of a passed in id. It comes in through the mechanism of Express. Specifying the path like this will have Express create a property of `req.params.id`. What you need to do is to verify that the request for a deletion of a user is actually the id that exists in the token. This way a user cannot delete an account that does not belong to them.

Look at session.js again and you see where the mongodb _id property of the retrieved document is captured. This is what is going to be passed back in the Rest request URL path portion to identify a user.

The middleware function `authHelper.checkAuth` is injected to do the verification that a valid token exists for the request. That middleware-injected function will return an error if the token is not acceptable and then the route function will never get called.

If everything proceeds correctly, the route function executes, and the document is removed from your collection, and the user account is gone. `findOneAndDelete()` is used as there would be one and only one document with that _id.

There is a helper function from the mongodb module to take the string and get it into the proper form needed. Here is the code for the user deletion handler:

## PART II: The Service Layer (Node.js)

```
router.delete('/:id', authHelper.checkAuth, function (req, res, next) {
  // Verify that the passed in id to delete is the same as that in the
  // auth token
  if (req.params.id != req.auth.userId)
    return next(new Error('Invalid request for account deletion'));

  // MongoDB should do the work of queuing this up and retrying if there is
  // a conflict, According to their documentation.
  // This requires a write lock on their part.
  req.db.collection.findOneAndDelete(
  { type: 'USER_TYPE', _id: ObjectId(req.auth.userId) },
  function (err, result) {
    if (err) {
      console.log("POSSIBLE USER DELETION CONTENTION? err:", err);
      return next(err);
    } else if (result.ok != 1) {
      console.log("POSSIBLE USER DELETION ERROR? result:", result);
      return next(new Error('Account deletion failure'));
    }

    res.status(200).json({ msg: "User Deleted" });
  });
});
```

### GET /api/users/:id

This route handler retrieves a single user by their id. The app would have already called to get a session token and then have access to the id of the user to pass it into this API call to retrieve the user document. Since you actually have the `_id` (object id), you can retrieve the document faster than if you had to query for it some other way. There is always an index created for the `_id` property.

The retrieval is done and populated with a transfer object. Notice that you are also tweaking the HTTP header for the response. That is necessary in order to stop caching from happening. Otherwise, when you get to the UI presentation code and are trying to retrieve a user, you might not get the most up-to-date one. Here is the code for the route handler to get a single user by their id:

```
router.get('/:id', authHelper.checkAuth, function (req, res, next) {
  // Verify that the passed in id to delete is the same the auth token
  if (req.params.id != req.auth.userId)
    return next(new Error('Invalid request for account fetch'));

  req.db.collection.findOne({ type: 'USER_TYPE',
                              _id: ObjectId(req.auth.userId) },
  function (err, doc) {
    if (err) return next(err);

    var xferProfile = {
      email: doc.email,
```

```
      displayName: doc.displayName,
      date: doc.date,
      settings: doc.settings,
      newsFilters: doc.newsFilters,
      savedStories: doc.savedStories
    };
    res.header("Cache-Control", "no-cache, no-store, must-revalidate");
    res.header("Pragma", "no-cache");
    res.header("Expires", 0);
    res.status(200).json(xferProfile);
  });
});
```

**PUT /api/users/:id**
A put is used to update a user, such as in the case where they have altered thier news filters. The code is very similar to what you needed for the initial post of the user, except you now need to worry about a conflict happening upon a database write operation. Here is the code for the user update handler:

```
router.put('/:id', authHelper.checkAuth, function (req, res, next) {
  // Verify that the passed in id is the same as that in the auth token
  if (req.params.id != req.auth.userId)
    return next(new Error('Invalid request for account deletion'));

  // Limit the number of newsFilters
  if (req.body.newsFilters.length > process.env.MAX_FILTERS)
    return next(new Error('Too many news newsFilters'));

  // clear out leading and trailing spaces
  for (var i = 0; i < req.body.newsFilters.length; i++) {
    if ("keyWords" in req.body.newsFilters[i] &&
        req.body.newsFilters[i].keyWords[0] != "") {
      for (var j = 0; j < req.body.newsFilters[i].keyWords.length; j++) {
        req.body.newsFilters[i].keyWords[j] =
          req.body.newsFilters[i].keyWords[j].trim();
      }
    }
  }

  // Validate the newsFilters
  var schema = {
    name: joi.string().min(1).max(30).regex(/^[-_ a-zA-Z0-9]+$/).required(),
    keyWords: joi.array().max(10).items(joi.string().max(20)).required(),
    enableAlert: joi.boolean(),
    alertFrequency: joi.number().min(0),
    enableAutoDelete: joi.boolean(),
    deleteTime: joi.date(),
    timeOfLastScan: joi.date(),
    newsStories: joi.array(),
    keywordsStr: joi.string().min(1).max(100)
```

## PART II: The Service Layer (Node.js)

```
    };

    async.eachSeries(req.body.newsFilters, function (filter, innercallback) {
      joi.validate(filter, schema, function (err) {
        innercallback(err);
      });
    }, function (err) {
      if (err) {
        return next(err);
      } else {
        // MongoDB implements optimistic concurrency for us.
        // We were not holding on to the document anyway, so we just do a
        // quick read and replace of just those properties and not the
        // complete document.
        // It matters if news stories were updated in the mean time (i.e.
        // user sat there taking their time updating their news profile)
        // because we will force that to update as part of this operation.
        // We need the {returnOriginal: false}, so a test could verify what
        // happened, otherwise the default is to return the original.
        req.db.collection.findOneAndUpdate({ type: 'USER_TYPE',
                                             _id: ObjectId(req.auth.userId) },
                    { $set: { settings: { requireWIFI: req.body.requireWIFI,
                       enableAlerts: req.body.enableAlerts },
                       newsFilters: req.body.newsFilters } },
          { returnOriginal: false },
          function (err, result) {
            if (err) {
              console.log("+++POSSIBLE USER PUT CONTENTION ERROR?+++ err:",
                          err);
              return next(err);
            } else if (result.ok != 1) {
              console.log("+++POSSIBLE CONTENTION ERROR?+++ result:",
                          result);
              return next(new Error('User PUT failure'));
            }

            req.node2.send({ msg: 'REFRESH_STORIES', doc: result.value });
            res.status(200).json(result.value);
          });
      }
    });
});
```

Notice the code to limit the news filter size. You will put code in the UI to limit that as well, but that could be hacked in the browser, or by someone sending a bogus `put` request. You have to guard against potential tampering as you would otherwise have a crash, or at least a failure of the MongoDB update.

There is the use of this amazingly useful module called async. This allows calls to the joi library to happen over and over and allows waiting for each callback to return for each of the filters for a user. When all are processed, the final anonymous function is called.

## Chapter 12: NewsWatcher App Development

The `$set` operation is used to update only individual properties and not the entire document.

There is some error checking code in there to detect any contention error. I had tried over and over and have never seen one happen.

**POST /api/users/:id/savedstories**
In the user document, there is an array used for saving stories that a user wants to keep around. This route will take the JSON of the passed-in request body as the story to save. The id in the route is the id of the user that is requesting the saving of the story.

There are a few checks that need to happen before saving a story. You need to verify that the story is not already inserted. There is also a limit on the number of stories that can be saved, so that has to be checked also.

Stories have an id associated with them to be able to identify them in cases like this where you don't want duplicates saved or shared. You will later see the code that creates that id. Here is the code for posting a story to be saved:

```
router.post('/:id/savedstories', authHelper.checkAuth, function (req, res,
           next) {
  // Verify that the id to delete is the same as in the auth token
  if (req.params.id != req.auth.userId)
    return next(new Error('Invalid request for saving story'));

  // Validate the body
  var schema = {
    contentSnippet: joi.string().max(200).required(),
    date: joi.date().required(),
    hours: joi.string().max(20),
    imageUrl: joi.string().max(300).required(),
    keep: joi.boolean().required(),
    link: joi.string().max(300).required(),
    source: joi.string().max(50).required(),
    storyID: joi.string().max(100).required(),
    title: joi.string().max(200).required()
  };

  joi.validate(req.body, schema, function (err) {
    if (err)
      return next(err);

    // make sure:
    // A. Story is not already in there.
    // B. We limit the number of saved stories to 30
    // Not allowed at free tier!!!
    // req.db.collection.findOneAndUpdate({ type: 'USER_TYPE',
    // _id: ObjectId(req.auth.userId),
    // $where: 'this.savedStories.length<29' },
    req.db.collection.findOneAndUpdate({ type: 'USER_TYPE',
```

## PART II: The Service Layer (Node.js)

```
                        _id: ObjectId(req.auth.userId) },
    { $addToSet: { savedStories: req.body } },
    { returnOriginal: true },
    function (err, result) {
      if (result && result.value == null) {
        return next(new Error('Over the save limit,
                         or story already saved'));
      } else if (err) {
        console.log("+++POSSIBLE CONTENTION ERROR?+++ err:", err);
        return next(err);
      } else if (result.ok != 1) {
        console.log("+++POSSIBLE CONTENTION ERROR?+++ result:", result);
        return next(new Error('Story save failure'));
      }

      res.status(200).json(result.value);
    });
  });
});
```

*Note:* *The $where above does not work with the free tier of Atlas and must be omitted.*

### DELETE /api/users/:id/savedstories/:sid

This is similar to the other functions and accomplishes the verification of the story existing, before being able to delete it. The `$pull` operator is used with the array property of the document to delete the entry. Here is the code for deleting a saved story:

```
router.delete('/:id/savedstories/:sid', authHelper.checkAuth, function (req, res, next) {
  if (req.params.id != req.auth.userId)
    return next(new Error('Invalid request for deletion of saved story'));

  req.db.collection.findOneAndUpdate({ type: 'USER_TYPE',
                        _id: ObjectId(req.auth.userId) },
    { $pull: { savedStories: { storyID: req.params.sid } } },
    { returnOriginal: true },
    function (err, result) {
      if (err) {
        console.log("+++POSSIBLE CONTENTION ERROR?+++ err:", err);
        return next(err);
      } else if (result.ok != 1) {
        console.log("+++POSSIBLE CONTENTION ERROR?+++ result:", result);
        return next(new Error('Story delete failure'));
      }

      res.status(200).json(result.value);
    });
});
```

Chapter 12: NewsWatcher App Development

# 12.12 Home News Routing (routes/homeNews.js)

Home news stories are those stories that are visible when the application UI is first seen by the user. There is only one single route needed for the homeNews resource. A user does not need to be logged in to see these stories.

| Verb and Path | Result |
|---|---|
| GET /api/homenews | Get all of the top news stories |

At the top and bottom of the file is the usual code to expose the module as shown here:

```
"use strict";
var express = require('express');

var router = express.Router();

...this part left out...

module.exports = router;
```

Here is the route handler for setting the news stories for the home page.

**GET /api/homenews**
Retrieving all top news stories is done by directly getting the array that holds them from our one single document for that purpose. They are the same for all users.

```
router.get('/', function (req, res, next) {
  req.db.collection.findOne({ _id: process.env.GLOBAL_STORIES_ID }, { homeNewsStories: 1 }, function (err, doc) {
    if (err)
      return next(err);
    res.status(200).json(doc.homeNewsStories);
  });
});
```

The array also works great, as you can send that back in the response and then React can bind it on the client side.

# 12.13 Shared News Routing (routes/sharedNews.js)

Shared news stories are those that are seen by all users. People can save, view and comment on news stories. Here are the specific routes for the sharedNews resource:

| Verb and Path | Result |
| --- | --- |
| POST /api/sharednews | Share a news story as contained in the JSON body |
| GET /api/sharednews | Get all of the shares news stories |
| DELETE /api/sharednews/:sid | Delete a news story that has been shared. |
| POST /api/sharednews/:sid/comments | Add a comment to a specified shared news story |

At the top and bottom of the file is the usual code as shown here:

```
"use strict";
var express = require('express');
var joi = require('joi'); // For data validation
var authHelper = require('./authHelper');

var router = express.Router();

...this part left out...

module.exports = router;
```

Here are each of the route path handlers.

**POST /api/sharednews**
This code is very similar to the code you already saw for saving a story in user.js. The only difference is that now, the story is being copied into a different document where all NewsWatcher users can view stories and comment on them.

There is a limit set for the number of possible shared stories. There is also a test to make sure the story was not already shared. If all looks good, the document is created. Here is the code for sharing a story:

```
router.post('/', authHelper.checkAuth, function (req, res, next) {
  // Validate the body
  var schema = {
    contentSnippet: joi.string().max(200).required(),
    date: joi.date().required(),
```

## Chapter 12: NewsWatcher App Development

```
    hours: joi.string().max(20),
    imageUrl: joi.string().max(300).required(),
    keep: joi.boolean().required(),
    link: joi.string().max(300).required(),
    source: joi.string().max(50).required(),
    storyID: joi.string().max(100).required(),
    title: joi.string().max(200).required()
  };

  joi.validate(req.body, schema, function (err) {
    if (err) return next(err);
    // We first make sure we are not at the count limit.
    req.db.collection.count({ type: 'SHAREDSTORY_TYPE' },
                      function (err, count)
    {
      if (err) return next(err);
      if (count > process.env.MAX_SHARED_STORIES)
        return next(new Error('Shared story limit reached'));

      // Make sure the story was not already shared
      req.db.collection.count({ type: 'SHAREDSTORY_TYPE',
        _id: req.body.storyID }, function (err, count)
      {
        if (err) return next(err);
        if (count > 0)
          return next(new Error('Story was already shared.'));

        // We set the id and guarantee uniqueness or failure happens
        var xferStory = {
          _id: req.body.storyID,
          type: 'SHAREDSTORY_TYPE',
          story: req.body,
          comments: [{
            displayName: req.auth.displayName,
            userId: req.auth.userId,
            dateTime: Date.now(),
            comment: req.auth.displayName +
                    " thought everyone might enjoy this!"
          }]
        };

        req.db.collection.insertOne(xferStory,
          function createUser(err, result)
        {
          if (err)
            return next(err);

          res.status(201).json(result.ops[0]);
        });
      });
    });
  });
});
```

PART II: The Service Layer (Node.js)

### GET /api/sharednews
Retrieving all shared stories is done by directly getting the documents of type `SHAREDSTORY_TYPE` as follows:

```
router.get('/', authHelper.checkAuth, function (req, res, next) {
  req.db.collection.find({ type: 'SHAREDSTORY_TYPE' }).toArray(function (err, docs) {
    if (err)
      return next(err);

    res.status(200).json(docs);
  });
});
```

You know there will not be more than 30 so it is ok to have an array returned and not use a cursor to iterate through the results. The array type works great, as you can send that back in the response and then React can bind it on the client side.

### DELETE /api/sharednews/:sid
Individual shared stories can be deleted. You will not actually be calling this from the presentation layer, but need it just for testing purposes to clean up after yourself. It can either be commented out or have some checks done to only allow an admin account to call it. The code is as follows:

```
router.delete('/:sid', authHelper.checkAuth, function (req, res, next) {
  req.db.collection.findOneAndDelete({ type: 'SHAREDSTORY_TYPE',
    _id: req.params.sid }, function (err, result)
  {
    if (err) {
      console.log("+++POSSIBLE CONTENTION ERROR?+++ err:", err);
      return next(err);
    } else if (result.ok != 1) {
      console.log("+++POSSIBLE CONTENTION ERROR?+++ result:", result);
      return next(new Error('Shared story deletion failure'));
    }

    res.status(200).json({ msg: "Shared story Deleted" });
  });
});
```

### POST /api/sharednews/:sid/comments
To add a comment, you need the id of the story and the JSON body for the comment. Since you have a partially normalized design here with separate documents for each story, you will not have as much write contention that way. There will still be concurrent access issues for each individual story as multiple comment additions will possibly conflict. This is why the `findOneAndUpdate()` call is used, as it will handle this for you.

## Chapter 12: NewsWatcher App Development

Notice that this can fail if there are already 30 comments added. There are three different parts to the query criteria used. The first two narrow it down to exactly what is being searched for. Then the $where operator is used and the actual JavaScript object is accessed to check the array length. The shared story comment is added as follows:

```
router.post('/:sid/Comments', authHelper.checkAuth, function (req, res,
next) {
  // Validate the body
  var schema = {
    comment: joi.string().max(250).required()
  };

  joi.validate(req.body, schema, function (err) {
    if (err)
      return next(err);

    var xferComment = {
      displayName: req.auth.displayName,
      userId: req.auth.userId,
      dateTime: Date.now(),
      comment: req.body.comment.substring(0, 250)
    };

    // Not allowed at free tier!!!req.db.collection.findOneAndUpdate({
    // type: 'SHAREDSTORY_TYPE', _id: req.params.sid, $where:
    // 'this.comments.length<29' },
    req.db.collection.findOneAndUpdate({ type: 'SHAREDSTORY_TYPE',
                                         _id: req.params.sid },
      { $push: { comments: xferComment } },
      function (err, result) {
        if (result && result.value == null) {
          return next(new Error('Comment limit reached'));
        } else if (err) {
          console.log("+++POSSIBLE CONTENTION ERROR?+++ err:", err);
          return next(err);
        } else if (result.ok != 1) {
          console.log("+++POSSIBLE CONTENTION ERROR?+++ result:", result);
          return next(new Error('Comment save failure'));
        }

        res.status(201).json({ msg: "Comment added" });
      });
  });
});
```

***Note:*** *The $where above does not work with the free tier of Atlas and must be omitted.*

PART II: The Service Layer (Node.js)

# 12.14 Forked Node Process (app_FORK.js)

You never want to have any compute-intensive code running in your main Node.js process. If you do, it will overwhelm the V8 JavaScript processing thread and your web service will become unresponsive. There are reasonable solutions to this, such as forking off other processes from your main process and having code execute there. Remember also, that I said it would probably be better to use AWS Lambda for this type of processing.

In order to architect your application correctly, you need to consider what needs to be moved off to the secondary processes. In the case of NewsWatcher, you can identify a few pieces of code that really need to be sent off to be run on a second Node.js process that is waiting to do any processing.

You can create a file named app_FORK.js and put a few functions of code in there. One section of code would be that which is periodically run on a timer for any batch type of work. For example, you need to populate the master news document with the latest news stories every once in a while and then run something to match all of the filters of the users.

Other code in app_Fork.js signaled to run by sending a message to this second Node process from the main process. For example, if a user ever alters their filters, code runs to update the stories that match for that single user.

Let's start with the top of the app_FORK.js file and look at the initialization code. The first lines will set up what is required for your module usage. One thing to note is that the database connection cannot be shared across processes, so you need to establish that again here.

```
"use strict";
var bcrypt = require('bcryptjs');
var https = require("https");
var async = require('async');
var assert = require('assert');
var ObjectId = require('mongodb').ObjectID;
var MongoClient = require('mongodb').MongoClient;

var globalNewsDoc;
const NEWYORKTIMES_CATEGORIES = ["home", "world", "national", "business", "technology"];

//
// MongoDB database connection initialization
//
var db = {};
```

## Chapter 12: NewsWatcher App Development

```
MongoClient.connect(process.env.MONGODB_CONNECT_URL, function (err, client)
{
  assert.equal(null, err);
  db.client = client;
  db.collection = client.db('newswatcherdb').collection('newswatcher');
  console.log("Fork is connected to MongoDB server");
});
```

Let's look at how you communicate back and forth between Node processes. There is a global variable made available in Node named `process`. It is used for accessing process-related properties and functions. One of those functions is `send()`. It can be used to send messages back to the parent process that forked the process. The `on()` function is for handling messages sent to this forked process from the main process.

```
process.on('message', function (m) {
  if (m.msg) {
    if (m.msg == 'REFRESH_STORIES') {
      setImmediate(function (doc) {
        refreshStoriesMSG(doc, null);
      }, m.doc);
    }
  } else {
    console.log('Message from master:', m);
  }
});
```

The one message sent from the main process is to handle changes to a user's filter. You schedule the handling of that and return immediately from the event. It needs to be delayed, or queued up as you could have other scheduled timer functions firing off, as well as other message requests coming from different users. The `setImmediate()` function is used to set up callbacks to run after I/O handling in this second node process.

**Refresh of a user's filters**

You can now look at the function that reacts to a user who has just updated their filters. This would be the code that is run in response to your message handling. The basic algorithm loops through each filter of a user. For each filter, the code checks if there are any stories in the master news list that match the keywords. There is a limit to the number of matching stories.

When the update of the user document happens, the `$set` operator is used and only a single property of the document is updated. Just the array property that holds the news filters.

There is code in here for testing purposes. This is used to be able to verify you have a known news story to test for a special predetermined keyword string. You will later see how this is used when the test code is discussed. Here is the code for user story matching:

## PART II: The Service Layer (Node.js)

```
function refreshStoriesMSG(doc, callback) {
  if (!globalNewsDoc) {
    db.collection.findOne({ _id: process.env.GLOBAL_STORIES_ID },
                    function (err, gDoc)
    {
      if (err) {
        console.log('FORK_ERROR: readDocument() read err:' + err);
        if (callback)
          return callback(err);
        else
          return;
      } else {
        globalNewsDoc = gDoc;
        refreshStories(doc, callback);
      }
    });
  } else {
    refreshStories(doc, callback);
  }
}

function refreshStories(doc, callback) {
  // Loop through all newsFilters and seek matches for all returned stories
  for (var filterIdx = 0; filterIdx < doc.newsFilters.length; filterIdx++)
  {
    doc.newsFilters[filterIdx].newsStories = [];
    for (var i = 0; i < globalNewsDoc.newsStories.length; i++) {
      globalNewsDoc.newsStories[i].keep = false;
    }

    // If there are keyWords, then filter by them
    if ("keyWords" in doc.newsFilters[filterIdx] &&
        doc.newsFilters[filterIdx].keyWords[0] != "") {
      var storiesMatched = 0;
      for (var i=0; i < doc.newsFilters[filterIdx].keyWords.length; i++) {
        for (var j=0; j < globalNewsDoc.newsStories.length; j++) {
          if (globalNewsDoc.newsStories[j].keep == false) {
            var s1 = globalNewsDoc.newsStories[j].title.toLowerCase();
            var s2 =
              globalNewsDoc.newsStories[j].contentSnippet.toLowerCase();
            var keyword =
              doc.newsFilters[filterIdx].keyWords[i].toLowerCase();
            if (s1.indexOf(keyword) >= 0 || s2.indexOf(keyword) >= 0) {
              globalNewsDoc.newsStories[j].keep = true;
              storiesMatched++;
            }
          }
          if (storiesMatched == process.env.MAX_FILTER_STORIES)
            break;
        }
        if (storiesMatched == process.env.MAX_FILTER_STORIES)
          break;
      }
```

```
      for (var k = 0; k < globalNewsDoc.newsStories.length; k++) {
        if (globalNewsDoc.newsStories[k].keep == true) {
          doc.newsFilters[filterIdx].newsStories.
            push(globalNewsDoc.newsStories[k]);
        }
      }
    }
  }

  // For the test runs, we can inject news stories under our control.
  if (doc.newsFilters.length == 1 &&
    doc.newsFilters[0].keyWords.length == 1
    && doc.newsFilters[0].keyWords[0] == "testingKeyword") {
    for (var i = 0; i < 5; i++) {
      doc.newsFilters[0].newsStories.push(globalNewsDoc.newsStories[0]);
      doc.newsFilters[0].newsStories[0].title = "testingKeyword title" + i;
    }
  }

  // Do the replacement of the news stories
  db.collection.findOneAndUpdate({ _id: ObjectId(doc._id) }, { $set: {
"newsFilters": doc.newsFilters } }, function (err, result) {
    if (err) {
      console.log('FORK_ERROR Replace of newsStories failed:', err);
    } else if (result.ok != 1) {
      console.log('FORK_ERROR Replace of newsStories failed:', result);
    } else {
      if (doc.newsFilters.length > 0) {
        console.log({ msg: 'MASTERNEWS_UPDATE first filter news length = '
+ doc.newsFilters[0].newsStories.length });
      } else {
        console.log({ msg: 'MASTERNEWS_UPDATE no newsFilters' });
      }
    }
    if (callback)
      return callback(err);
  });
}
```

This second function is used in multiple places, so I pulled it out to be on its own.

**Timer event to populate the master news list**
Every few hours, a function runs that fetches all news stories from the source news service provider. This means that it works on a batch of data and could take some time before it finishes. This function was placed in Node process because it is long-running. It is still significant enough that it is better to place it there so as to not upset the main Node process CPU usage. This could be offloaded to a completely different machine through some other queuing mechanism.

# PART II: The Service Layer (Node.js)

The first thing that happens when the time interval fires, is the sending if an HTTP request to the news feed API provided by the New York Times API service. The results from that are placed into formatted news elements in the master document `newsStories` array. Notice the use of the async module to be able to loop a number of times and also set a .5 second delay between each batch news request from NYT. This is because there is a restriction with the usage such that you can't call it more than five times a second, or they may disable your IP address from accessing their API.

There is an id needed for each story. A GUID could have been generated, but the problem is that the same story might appear again in the next batch of news and that would cause problems if you thought it was a new news story. A hash of the link will turn out to be the best way to uniquely identify a story. The link itself would be unusable later in a URL to pass in as an ID, but the hash value works, as long as you replace certain characters of it.

Once all of the news stories are in place, you can go through all of the user documents and, for each user, do the story matching against the news filters. This is done in a somewhat tricky way. The `async.doWhilst()` functionality is used. This way, it can handle the difficulty of managing multiple async calls, one at a time, in a simple way. The code will keep running as long as there is processing to do.

You have to consider the throughput capability on the MongoDB side and not overwhelm it, so it is good to have this processing serialized. You need to keep plenty of headroom for your normal user interactions. Here is the code:

```
var count = 0;
newsPullBackgroundTimer = setInterval(function () {
  // The New York Times news service can't be called called more than five
  // times a second. We call it over and over again, because there are
  // multiple news category is, so space each out by half a second
  var date = new Date();
  console.log("app_FORK: datetime tick: " + date.toUTCString());
  async.timesSeries(NEWYORKTIMES_CATEGORIES.length, function (n, next) {
    setTimeout(function () {
      console.log('Get news stories from NYT. Pass #', n);
      try {
        https.get({
          host: 'api.nytimes.com',
          path: '/svc/topstories/v2/' + NEWYORKTIMES_CATEGORIES[n] +
            '.json',
          headers: { 'api-key': process.env.NEWYORKTIMES_API_KEY }
        }, function (res) {
          var body = '';
          res.on('data', function (d) {
            body += d;
          });
          res.on('end', function () {
            next(null, body);
```

## Chapter 12: NewsWatcher App Development

```
        });
      }).on('error', function (err) {
        // handle errors with the request itself
        console.log({ msg: 'FORK_ERROR', Error: err.message });
        return;
      });
    }
    catch (err) {
      count++;
      if (count == 3) {
        console.log('app_FORK.js: shuting down timer:' + err);
        clearInterval(newsPullBackgroundTimer);
        clearInterval(staleStoryDeleteBackgroundTimer);
        process.disconnect();
      }
      else {
        console.log('app_FORK.js error. err:' + err);
      }
    }
  }, 500);
}, function (err, results) {
  if (err) {
    console.log('failure');
  } else {
    console.log('success');
    // Do the replacement of the news stories in the single master doc
    db.collection.findOne({ _id: process.env.GLOBAL_STORIES_ID },
    function (err, gDoc) {
      if (err) {
        console.log({ msg: 'FORK_ERROR', Error: 'Error with the global
news doc read request: ' + JSON.stringify(err.body, null, 4) });
      } else {
        gDoc.newsStories = [];
        gDoc.homeNewsStories = [];
        var allNews = [];
        for (var i = 0; i < results.length; i++) {
          try {
            var news = JSON.parse(results[i]);
          } catch (e) {
            console.error(e);
            return;
          }
          for (var j = 0; j < news.results.length; j++) {
            var xferNewsStory = {
              link: news.results[j].url,
              title: news.results[j].title,
              contentSnippet: news.results[j].abstract,
              source: news.results[j].section,
              date: new Date(news.results[j].updated_date).getTime()
            };
            // Only take stories with images
            if (news.results[j].multimedia.length > 0) {
              xferNewsStory.imageUrl = news.results[j].multimedia[0].url;
```

217

```
            allNews.push(xferNewsStory);
            // Populate the home page stories
            if (i == 0) {
              gDoc.homeNewsStories.push(xferNewsStory);
            }
          }
        }
      }

      async.eachSeries(allNews, function (story, innercallback) {
        bcrypt.hash(story.link, 10, function getHash(err, hash) {
          if (err)
            innercallback(err);

          // Only add the story if it is not in there already.
          // Stories on NYT can be shared between categories
          story.storyID = hash.replace(/\+/g, '-').
                          replace(/\//g, '_').replace(/=+$/, '');
          if (gDoc.newsStories.findIndex(function (o) {
            if (o.storyID == story.storyID || o.title == story.title)
              return true;
            else
              return false;
          }) == -1) {
            gDoc.newsStories.push(story);
          }
          innercallback();
        });
      }, function (err) {
        if (err) {
          console.log('failure on story id creation');
        } else {
          console.log('story id creation success');
          globalNewsDoc = gDoc;
          setImmediate(function () {
            refreshAllUserStories();
          });
        }
      });
    }
  });
  }
});
}, 240 * 60 * 1000);

function refreshAllUserStories() {
  db.collection.findOneAndUpdate({ _id: globalNewsDoc._id },
    { $set: { newsStories: globalNewsDoc.newsStories,
              homeNewsStories: globalNewsDoc.homeNewsStories } },
    function (err, result)
  {
    if (err) {
      console.log('FORK_ERROR Replace of global newsStories failed:', err);
```

```
      } else if (result.ok != 1) {
        console.log('Replace of global newsStories failed:', result);
      } else {
        // For each NewsWatcher user, do news matching on their newsFilters
        var cursor = db.collection.find({ type: 'USER_TYPE' });
        var keepProcessing = true;
        async.doWhilst(
          function (callback) {
            cursor.next(function (err, doc) {
              if (doc) {
                refreshStories(doc, function (err) {
                  callback(null);
                });
              } else {
                keepProcessing = false;
                callback(null);
              }
            });
          },
          function () { return keepProcessing; },
          function (err) {
            console.log('Timer: Refreshed and matched. err:', err);
          });
      }
    });
}
```

*Note: Anytime you use these async functions, you must be really careful of how they operate with their async and sync capabilities and make sure you call the required callbacks correctly in the right place. Error handling can also be a bit tricky here.*

**Deleting old stories**
There needs to be code that will delete shared stories after a certain amount of time. This becomes another timer that goes off periodically to do this processing. The code is as follows:

```
staleStoryDeleteBackgroundTimer = setInterval(function () {
  db.collection.find({ type: 'SHAREDSTORY_TYPE' }).toArray(
  function (err, docs) {
    if (err) {
      console.log('Fork could not get shared stories. err:', err);
      return;
    }

    async.eachSeries(docs, function (story, innercallback) {
      // Go off the date of the time the story was shared
      var d1 = story.comments[0].dateTime;
      var d2 = Date.now();
      var diff = Math.floor((d2 - d1) / 3600000);
      if (diff > 72) {
        db.collection.findOneAndDelete({ type: 'SHAREDSTORY_TYPE',
                                    _id: story._id },
```

```
        function (err, result) {
          innercallback(err);
        });
      } else {
        innercallback();
      }
    }, function (err) {
      if (err) {
        console.log('stale story deletion failure');
      } else {
        console.log('stale story deletion success');
      }
    });
  });
}, 24 * 60 * 60 * 1000);
```

# 12.15 Securing with HTTPS

At this point, I need to discuss an important security measure that needs to be put into place. You just can't host your REST API endpoint without encrypting traffic back and forth. You fix this by only permitting HTTPS connections. That way all traffic is encrypted and signed so as to be tamper-resistant and harder to eavesdrop on.

Since the Node.js service is exposed through the Elastic Beanstalk app, you don't need to make any changes to your Node.js code to secure it, it is all done through AWS. If your Node.js instance was exposed directly to the Internet and serving up the traffic directly, then you would do some simple configuration on the Node.js side to install a certificate and key files and then make a few modifications to the Node code.

In this case, Elastic Beanstalk keeps Node from being directly exposed. The Elastic Beanstalk service acts as a reverse proxy and that is where you need to set up SSL. You will need to get your own domain name and install a certificate for your Elastic Beanstalk service.

*Note:* When you launch your application through VS Code, it will not accept HTTPS locally on your machine. However, tests run against a production deployment must be altered to use HTTPS. Your test code needs to make the appropriate changes to the URL being tested.

**Securing communications to MongoDB**
For performance reasons, you will want to place your database in the same AWS datacenter as your Elastic Beanstalk Node app. As an added bonus of doing this, the ability to secure the communication between your Node.js service and your database becomes easier. This is because everything is sent over the internal data center network and never gets out over the

Chapter 12: NewsWatcher App Development

public Internet. As a further measure of security, you can also communicate over an SSL connection if you use a dedicated plan from MongoDB Inc.

With a dedicated plan, you also define custom firewall rules so that your database access is limited to specific IP address ranges and/or to specific AWS EC2 security groups.

# DNS and certificate setup

In the Introduction, you saw a physical topology diagram (see figure 4) that showed a domain name being routed through a DNS server that had a certificate to enable HTTPS. You can now go about setting that up.

First, go to the Route 53 service management console. Click **Register Domain** to start the process of getting your own domain, or you may need to click **Get started now** under the heading to do a domain registration, then click **Register Domain**. If you already have your own domain name and want to use it, you can do that but that is not covered in this book. Here is the Route 53 management console:

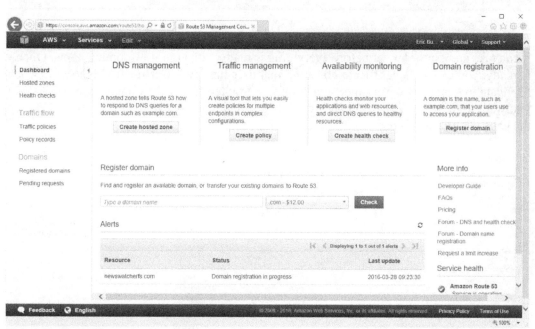

*Figure 62 - Route 53 service management console*

You can type in different names to try and you will be able to find one that is available. There is an initial charge as well as a small recurring fee when you purchase a domain name. For this implementation of the NewsWatcher sample app, I settled on newswatcher2rweb.com since newswatcher.com was already taken.

221

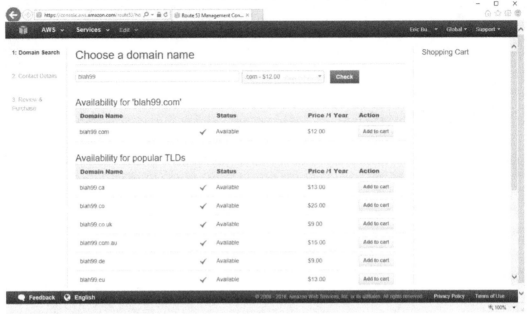

*Figure 63 - Route 53 domain availability*

It can take an hour or longer before everything is ready with your new domain name. Once it is ready, you can get a certificate all set up that uses the AWS Certificate Manager Service. Navigate to **Certificate Manager** and click **Get Started**. Enter the domain name variations you want to be supported. You can use wildcards and have the flexibility you need. Here is the page where that happens:

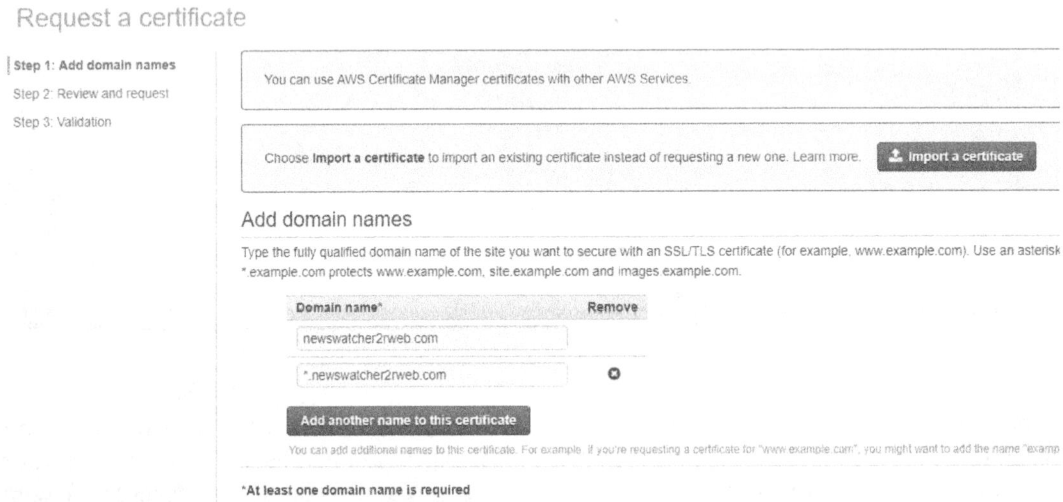

*Figure 64 - Add domain names to AWS Certificate Manager*

Chapter 12: NewsWatcher App Development

After this screen, there are a few more to click through. At some point, you will also need to validate this action through an email you will receive that verifies that you are the owner of the domain name that the certificate is being set up for.

A certificate can now be set on the Elastic Beanstalk load balancer. You do this through the AWS management portal. Here are the steps:

1. Open your AWS Elastic Beanstalk environment and on the left, find and select **Configuration**.
2. Click the gear by **Scaling** and change **Environment type** to be **Load balancing, auto scaling**.
3. Wait for this change to take effect and then click **Configuration** again from the selections on the left and then click **Load Balancing**.
4. Select your certificate from the **SSL certificate ID** drop down.
5. Click **Apply** at the bottom of the page.

Now you can watch the status there until it indicates that the configuration change is successful.

Recent Events

| Time | Type | Details |
| --- | --- | --- |
| 2016-04-07 13:46:48 UTC-0700 | INFO | Environment update completed successfully. |
| 2016-04-07 13:46:48 UTC-0700 | INFO | Successfully deployed new configuration to environment. |
| 2016-04-07 13:45:01 UTC-0700 | INFO | Updating environment newswatcher's configuration settings. |
| 2016-04-07 13:44:48 UTC-0700 | INFO | Environment update is starting. |

*Figure 65 - Elastic Beanstalk status*

Now you can set up the DNS routing to your Elastic Beanstalk load balancer. Go back into the Route 53 management console and click under the **DNS management** the **Hosted zones** link. Then click on your Domain name. Click **Create Record Set**.

223

PART II: The Service Layer (Node.js)

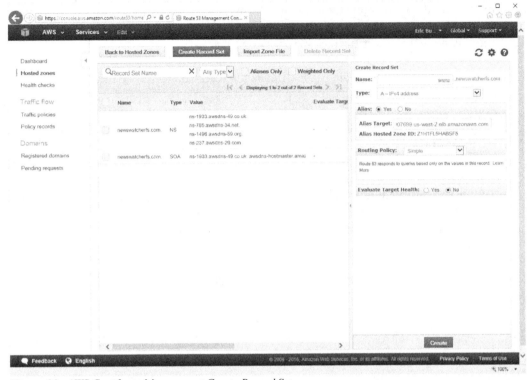

*Figure 66 - AWS Certificate Management Create Record Set*

Fill out the form on the right as shown in figure 69. For **Alias Target**, select your load balancer. You can add a record set with name "www" and then make one with a blank name.

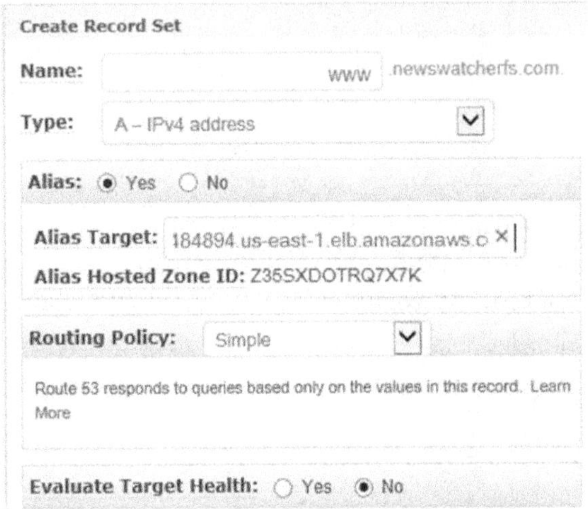

*Figure 67 - Route 53 create record set*

224

# Chapter 12: NewsWatcher App Development

You are almost done. You need to turn off HTTP access at the load balancer. To do that:
1. Go to the EC2 service management console and click **1 Load Balancers**.
2. Click the **Listeners** tab, then click **Edit**.
3. Delete the HTTP entry, then click **Save**.

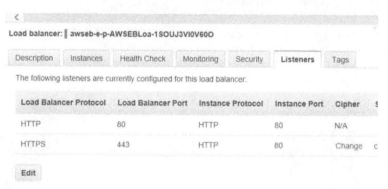

*Figure 68 - EC2 service Listeners tab*

Back at the Elastic Beanstalk environment page, if you click on the URL at the top, it will no longer work. In your browser, change the URL to start with `https:` and you will be told the certificate does not match. Now type in the URL to the domain name you registered and you will see everything up and working. HTTPS is now working. If you try, https://www.newswatcher2rweb.com/ it will work, but http://www.newswatcher2rweb.com/ will not work.

# 12.16 Deployment

At this point, you have everything in place to start trying out your middle-tier web service API that is implemented as an HTTP/Rest endpoint. You obviously would not build a service like this without testing it along the way. All of the discussion involving the testing of the service is discussed in the next chapter. You can see it deployed at this point and then understand the testing that is necessary.

Don't get the wrong impression. I certainly wrote the code in small iterations and tested each and every bit of it along the way. Good developers iterate and test everything as they go.

At this point, you can zip up your code and deploy it up to AWS. Before doing so, you can test things out with some tests that will be described in chapter 13. You will actually want to test on your local machine as well as deploy to some known staging cloud location and test there as well. You would be running the test code from your local machine to go against a staging site that is hosted in AWS.

# PART II: The Service Layer (Node.js)

Here are the folders and files I selected to have zipped up on my Windows machine. There is a right-click selection to create a zip file under **Send to** on a windows machine.

The build and public folders are not actually there yet, so leave those out. They are later folders that get added that hold the React application.

You follow the same instructions detailed previously to get your application code up and running. And now at this point have secured the traffic through HTTPS.

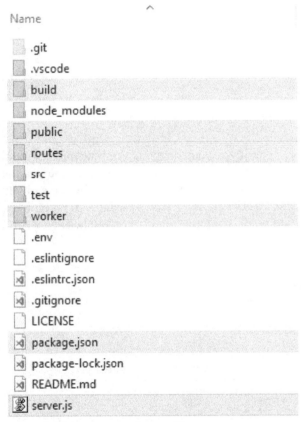

*Figure 69 - Selections to make a zip file*

**Note:** *If you find errors upon deployment, you can click on the **logs** selection in the UI of the Elastic Beanstalk AWS console. From the actions menu, select to download the full logs. In the folder you download you will find a file named eb-activity.log. Scroll and see if there are any indications as to why the npm install command failed. There seems to be an ongoing issue with AWS Elastic Beanstalk installs where they run out of memory when the install happens. To work around this, switch to a t2.small machine and then try your install again.*

# Chapter 13: Testing the NewsWatcher Rest API

Now comes the exciting part where you will get to see the HTTP/Rest Web API exercised. You will prove that the service is up and running locally, then test the deployment and verify everything in production. Once sufficiently proven, you can start the final task of creating a UI for NewsWatcher and be confident that the integration will go smoothly.

This chapter will present several practices that you will want to follow for exercising your code to fully test it. It is obviously a lot simpler to test and debug issues locally. Let's first look at how to employ techniques for debugging issues in production.

## 13.1 Debugging During Testing

Let's first talk about debugging techniques. You will need to do some of that as you run tests and need to examine the execution of your code.

In some cases, output logging to the Node console will provide you with enough clues to track down an issue. This means that you must log important things that are happening in the application. Beyond this, you will need a few tools to help you do your investigations.

One tool at your disposal is the VS Code debugger. Before deploying anything, you can run your code locally. You can use VS Code to debug your Node.js project code.

If you want to debug your Node.js code, you open your project, and launch the Node.js project by pressing F5. You can set up your breakpoints in advance, or add ones as needed that you want to be hit. Once your project is running, you run Mocha from the command line and exercise your code through tests you have written. Then you can step through your code.

*Note: There are some issues when forking a second process when running in debug mode. That is why I have the line you can uncomment that makes it possible. Uncomment "var node2 = cp.fork('./worker/app_FORK.js', [], { execArgv: ['--inspect=9229'] });".*

To set up debugging in VS code, click the debug icon. You will see a gear icon at the top of the window that you can click to create the launch.json file. This is the file that instructs VS Code how to proceed. You can set it up to have two configurations in the file. One will be for launching your node process with debugging capability. The other entry is for attaching to an already running Node process. When you click the gear icon, select the Node selection and create the file. Your file might look as follows:

# PART II: The Service Layer (Node.js)

```
{
  "version": "0.2.0",
  "configurations": [
    {
      "type": "node",
      "request": "attach",
      "name": "Attach to Remote",
      "address": "TCP/IP address of process to be debugged",
      "port": 9229,
      "localRoot": "${workspaceFolder}",
      "remoteRoot": "Absolute path to the remote directory containing the program"
    },
    {
      "type": "node",
      "request": "launch",
      "name": "Launch",
      "program": "${workspaceFolder}/server.js"
    },
    {
      "type": "node",
      "request": "attach",
      "name": "Attach",
      "port": 9229
    }
  ]
}
```

To the left of the gear is the start button (you can also press F5). The drop-down menu will show you the config section options that come from your launch.json file. The far right greater-than symbol in the box opens the output console window. It is a good idea to always have that open to view any statements or errors that get displayed.

*Figure 70 - Visual Studio Code debuggin*

You can open the launch.json file and look at it, but you do not need to make any adjustments to it as the defaults are just what you need, unless it is set to start node with another file. You

Chapter 13: Testing the NewsWatcher Rest API

can look up the documentation on the settings that can be used in the launch.json file on the Node website and VS Code Website.

To place a breakpoint, open a JavaScript file and click out to the left of the margin, or click on the line and press F9. Once you hit a breakpoint while running code, you get full access to inspect the call stack and variables.

# 13.2 Tools to Make an HTTP/Rest Call

You will probably want to use a tool to make individual calls to your API. That way, you can take small steps to get everything verified before you throw a test harness into the mix. I have installed Postman, Fiddler and Curl.exe on my machine for testing individual HTTP/Rest calls. There are many of these such tools. With these tools, you can send HTTP requests to your service and view the returned response.

From one of the recommended tools, you can set up the verb usage, as well as the headers and the JSON body content. You can do a send and then look at the returned response. Let's look at the UI of Fiddler and I will explain the basics of how to use it to call your API.

You can send an HTTP request to the route handler that registers a user. This is an obvious first place to start. To do that, you know you have the `app.use('/api/users', users)` call in your server.js file that takes you to the users.js function of `router.post('/', function(req, res, next) {}`. This is the code that creates a new registration.

You will want to make a post verb call and pass in a JSON body that contains the display name, email and password. You then expect a 201 return code to be given back to indicate a successful creation. In the return response will be the returned document that you could inspect.

You can now start up your node service with your project open and press F5 to debug, or press Ctrl+F5 to run without debugging. The first thing you notice is that you get a console app window that opens in the VS Code UI. That represents your node process running and the executing of your server.js file. You will see all of your console logging appear in this window. If you don't see it, click on the toolbar on the left to get to the debug area and you can open it from there.

Your Node.js server is running as a local process now. To make an HTTP request you can connect to localhost with the port number of 3000 and interact with the Rest API. You can open Fiddler and try this out. The first thing to do is to limit what traffic Fiddler sees going

# PART II: The Service Layer (Node.js)

back and forth. Otherwise, you will get lost in the stream of traffic. To do this and also make an HTTP request do the following:

1. In the Fiddle Web Debugger, click the **Filters** tab.
2. In the top section, check **Use Filters.**
3. In the **Request Headers** section, check **Show only if URL contains** and type in "localhost".

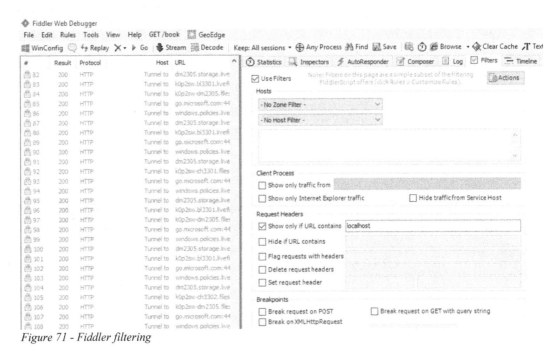

*Figure 71 - Fiddler filtering*

4. Click the drop-down menu **X -> Remove all** to get rid of all old traffic.

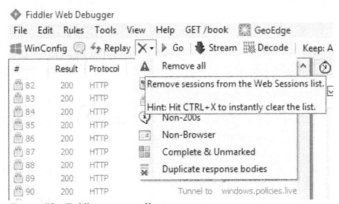

*Figure 72 - Fiddler remove all*

230

Chapter 13: Testing the NewsWatcher Rest API

5. Click the **Composer** tab, then click the **Options** sub-tab under that and make sure it looks as follow:

*Figure 73 - Options*

You want Fiddler to figure out the content header length value for you.

6. Click the **Raw** tab, and enter the following request:
```
POST http://localhost:3000/api/users HTTP/1.1
User-Agent: Fiddler
Host: localhost:3000
Content-Type: application/json
Content-Length: 85

{
"email" : "bush@sample.com",
"displayName" : "Bushman",
"password" : "abc1234#"
}
```

*Figure 74 - Fiddler RAW request*

231

# PART II: The Service Layer (Node.js)

7. Click **Execute**.

If you look in the left-hand pane, you will see the request being sent and a response returned. If you double-click the 201 response, you can examine it and see that it worked. It should look as follows if you click on **Raw** or **JSON** view:

*Figure 75 - Fiddler request and response*

You have now successfully seen your API exercised. Open the MongoDB Compass tool and you can also see that there is a new document created. In Compass, you will see something like the following:

232

Chapter 13: Testing the NewsWatcher Rest API

*Figure 76 - Atlas management portal with a new document*

You can now enter every single request through Fiddler to prove them all. Next, try to log in and get a token back:

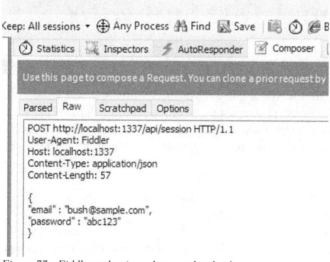

*Figure 77 - Fiddler to log in and get a token back*

233

PART II: The Service Layer (Node.js)

Look at the result and you will see a token in the response if you open it up. You will need this token to use in the next requests. You can next make a call to retrieve the user document and see if there are any news stories that have matched the filter.

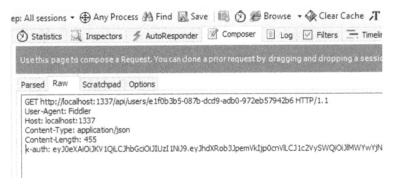

Figure 78 - Fiddler to see the token

In Fiddler, Click **Execute** and you will see a response with the returned user document and possibly some news stories. This is great progress, and you can continue to try all of your API calls and debug each one if needed. As you get each one working, you can create a test case for each.

As was mentioned, you might find a tool other than Fiddler that you like. Curl is a nice option because you just open a command prompt and execute it there. It has a simple syntax that you use to formulate your requests.

# 13.3 A Functional Test Suit with Mocha

Now you will take the next step of automating the testing of the API using a test suite. This will then become your functional test pass. You can start by implementing some of the same operations that were already tried when you used Fiddler. To start with, you need to add a new folder to your VS Code project to hold the tests. You can also set up the added node modules you need for the tests.

*Note: NPM lets you download different types of modules. Like previous modules I've shown you, you download them and then use a* `require` *statement to utilize them as code modules. Others, like Mocha, are not code modules, but are command-line tools. What you do is write a JavaScript file that Mocha will interpret and run. You then use a command-line to execute Mocha, and it does its work. Mocha is installed as a local part of your project. This means the executable will be referenced from that location. Mocha can also be installed globally if you like.*

# Chapter 13: Testing the NewsWatcher Rest API

Add the following to the package.json file:

```
"devDependencies": {
  "mocha": "^3.4.2",
  "supertest": "^3.0.0"
}
```

The `devDependencies` section is reserved for non-production modules that you will not need to deploy to a production build. They are only needed to run your test code locally.

We can now go through some of the files in the test folder to see how the NewsWatcher application can be tested. This first file will be used to hit the API endpoint to exercise the routes.

## Writing mocha tests (functional_api_crud.js)

You will make use of the supertest module and the assert module. The necessary `require()` calls can be set up for those. The server object also needs to be retrieved and given time to be initialized. There are special blocks of code that mocha will run before and after all tests. The rest of the code makes use of mocha test blocks to run tests with.

The request object is what you set up from the supertest module. With that, you make the HTTP `post` verb call and register a user.

You will use the `describe` keyword for any major test block and use the individual `it` keywords for each test inside of that. With just one test to run, it would look as follows:

```
var assert = require('assert');
var app = require('../server.js');
var request = require('supertest')(app);

describe('User cycle operations', function () {
  // Wait until the database is up and connected to.
  before(function (done) {
    setTimeout(function () {
       done();
    }, 5000);
  });

  // Shut everything down gracefully
  after(function (done) {
    app.db.client.close();
    app.node2.kill();
    app.close(done);
  });
```

```
    it("should create a new registered User", function (done) {
        request.post("/api/users")
        .send({
            email: 'bush@sample.com',
            displayName: 'Bushman',
            password: 'abc123*'
        })
        .end(function (err, res) {
            assert.equal(res.status, 201);
            assert.equal(res.body.displayName, "Bushman", "Name of user should be as set");
            done();
        });
    });
});
```

*Note:* Mocha actually supports several different styles of syntax, so don't be confused if you see other projects using Mocha and it does not look exactly like the code in this book.

I will go through the previous code to make sure you understand it. To start with, you have the usual Node.js require statements at the top. For the supertest module, you specify that you are going to hit the passed in web server endpoint of your Node project. The strings you set as parameters to `describe()` and `it()` are printed out for you as part of the test run. You should use text that helps you remember what you are testing.

With this code, you have a test that will verify that you can register a user. The `it` block takes a string to describe what you are testing and then a function to run. There is a `done()` function that you call to signal to mocha that it can move on to the next `it` block. These tests are each run sequentially. If you don't call `done()`, the test will eventually time out.

To use supertest, you specify a verb operation to use. This one is using `post`. You string together the function calls `send()` and `end()`. Each one will get called in sequence. The `send` function does the HTTP/Rest request and has the body set.

The `end()` function can get the response and validate the return code and values from the returned body. You use the assert module for validations.

There are some cases where you have a second test that relies on the results of the first test. This can be tricky if there is delayed processing of the first test code. You can either stack one test inside the other, or use a JavaScript `setTimeout()` call to delay your second test run by a bit and then have it run.

For example, when a user changes a news filters, that operation will return immediately. This means that the test code will move on to the next test. The way NewsWatcher works, is that it sends a message to the forked process to update the news stories. If you wanted to test a

## Chapter 13: Testing the NewsWatcher Rest API

change to a profile filter string, you would realize that it would take a second to do that in the background and you would put in a delay before the next code could ran.

The following code shows using a delay of three seconds before running a test:

```
it("should allow access if logged in ", function (done) {
   setTimeout(function () {
      request.get("/api/users/" + userId)
      .set('x-auth', token)
      .end(function (err, res) {
         assert.equal(res.status, 200);
         savedDoc = res.body.newsFilters[0].newsStories[0];
         console.log(JSON.stringify(savedDoc, null, 4));
         done();
      });
   }, 3000);
});
```

Here is a more complete set of tests that registers a user and then makes sure they can log in and then deletes the account to clean things up. The test right at the start is to verify that a person cannot log in if they don't first register.

There may be negative tests you can put in place to verify your error handling code. In this code below, I make use of local valuables inside the describe block to pass these between tests such as the token variable. For example, you need to capture the token at sign in time and keep using it on subsequent rest calls that you are testing.

```
var assert = require('assert');
var app = require('../server.js');
var request = require('supertest')(app);

describe('User cycle operations', function () {
   var token;
   var userId;
   var savedDoc;

   // Wait until the database is up and connected to.
   before(function (done) {
      setTimeout(function () {
         done();
      }, 3000);
   });

   // Shut everything down gracefully
   after(function (done) {
      app.db.client.close();
      app.node2.kill();
      app.close(done);
   });
```

```
   it("should deny unregistered user a login attempt", function (done) {
      request.post("/api/sessions").send({
         email: 'bush@sample.com',
         password: 'abc123*'
      })
        .end(function (err, res) {
         assert.equal(res.status, 500);
         done();
      });
   });

   it("should create a new registered User", function (done) {
      request.post("/api/users")
        .send({
         email: 'bush@sample.com',
         displayName: 'Bushman',
         password: 'abc123*'
      })
         .end(function (err, res) {
         assert.equal(res.status, 201);
         assert.equal(res.body.displayName, "Bushman", "Name of user should be as set");
         done();
      });
   });

   it("should not create a User twice", function (done) {
      request.post("/api/users")
        .send({
         email: 'bush@sample.com',
         displayName: 'Bushman',
         password: 'abc123*'
      })
         .end(function (err, res) {
         assert.equal(res.status, 500);
         assert.equal(res.body.message, "Error: Email account already registered", "Error should be already registered");
         done();
      });
   });

   it("should detect incorrect password", function (done) {
      request.post("/api/sessions")
        .send({
         email: 'bush@sample.com',
         password: 'wrong1*'
      })
         .end(function (err, res) {
         assert.equal(res.status, 500);
         assert.equal(res.body.message, "Error: Wrong password", "Error should be already registered");
         done();
      });   });
```

## Chapter 13: Testing the NewsWatcher Rest API

```javascript
    it("should allow registered user to login", function (done) {
       request.post("/api/sessions")
         .send({
           email: 'bush@sample.com',
           password: 'abc123*'
       })
         .end(function (err, res) {
           //<Session&Cookie code>cookies = res.headers['set-cookie'];
           token = res.body.token;
           userId = res.body.userId;
           assert.equal(res.status, 201);
           assert.equal(res.body.msg, "Authorized", "Message should be AUthorized");
           done();
         });
    });

    it("should allow registered user to logout", function (done) {
       request.del("/api/sessions/" + userId)
         .set('x-auth', token)
         .end(function (err, res) {
           assert.equal(res.status, 200);
           done();
         });
    });

    it("should not allow access if not logged in", function (done) {
       request.get("/api/users/" + userId)
         .end(function (err, res) {
           assert.equal(res.status, 500);
           done();
         });
    });

    it("should allow registered user to login", function (done) {
       request.post("/api/sessions")
         .send({
           email: 'bush@sample.com',
           password: 'abc123*'
       })
         .end(function (err, res) {
           token = res.body.token;
           userId = res.body.userId;
           assert.equal(res.status, 201);
           assert.equal(res.body.msg, "Authorized", " Authorized Message");
           done();
         });
    });
    it("should delete a registered User", function (done) {
       request.del("/api/users/" + userId)
         .set('x-auth', token)
         .end(function (err, res) {
           assert.equal(res.status, 200);
```

# PART II: The Service Layer (Node.js)

```
        done();
    });
  });
});
```

You run your Mocha test suite from a command prompt. On Windows machines you open a **Node.js command prompt**. You can also use GitHub Desktop and right click your repository and select **Open in Git shell**. Once you are at the command prompt, navigate to the location of your project if needed. This is not necessary if you open up a prompt from GitHub desktop.

You don't need to start up the node.js application, because the Mocha code will do that. You can run Mocha from the local project folder in the git shell window as follows:

```
./node_modules/.bin/mocha --timeout 30000 test/functional_api_crud.js
```

You may need to play around with the timeout argument. It is possible to get false failures because of Mocha timing-out and moving on to the next `it` block too quickly. The output for the complete functional test suite looks as follows:

*Figure 79 - Mocha functional test output*

Chapter 13: Testing the NewsWatcher Rest API

*Note*: *You should realize that, even if you are running against your local Node.js service, you are still hitting the real AWS hosted MongoDB database. You might need to open the Compass app and delete unwanted documents that you created through your tests. The tests included with the NewsWatcher sample are written to clean up after themselves. Of course, you could install a local copy of MongoDB on your machine and connect against that as well.*

To run against the deployed Node.js application in AWS, you can change the supertest usage to go against the production URL, such as https://www.newswatcher2rweb.com. You can see in the GitHub project that I have a few commented out lines for different ways to run the tests.

## 13.4 Performance and Load Testing

Writing an application that can serve a single user is not a big challenge. The real challenge comes when multiple people are all hitting the web service REST API at the same time.

To write the NewsWatcher sample app and get it to work for a single user was just two weeks of work for me. To get it to scale and handle the simulated load of many users, required much longer than that to work out all the issues. You certainly don't want to wait until your big production rollout to find that your code falls flat on its face when more than one person uses it.

How are you going to accomplish testing your code at scale? The only way to accomplish this is with a test suite that can provide usage in parallel and simulate multiple users.

There are UI testing tools that can record your usage of a website and replay it. They can be replayed more than once at the same time to simulate lots of interaction happening. This might work well for some sites that serve static content and have no concept of people logging in and exercising some unique workflow in the backend service layer.

Load testing enables you to prove the scaling of your application. You also use this to measure your SLA (Service Level Agreement) values under a constant load. You can experiment by increasing the load until you find the breaking point. This will tell you the absolute peak values you can run under. To do this, you need to alter the URL to be that of the production or cloud staging environment.

Running the load testing suite can also help you test out your Elastic Beanstalk scaling strategies and topology. It will of course be useful in verifying the performance of MongoDB that you have worked hard to optimize.

# 13.5 Running Lint

You can install the npm packages "eslint" and "eslint-plugin-react". These allow you to run validations that scan the code and look for syntax errors, uninitialized variables and even specific stying that you want. I would say that it is a requirement to have this in place. If you were writing in a language such as C++ that was compiled, you get that all done up front. Unfortunately, you write code with JavaScript, but then don't get to see errors until each line executes.

You will see that the package.json file has some scripts set up to run lint and also combine the running of lint with the running of the functional mocha tests. This is what is in the package.json file for that:

```
"scripts": {
  "lint": "eslint **/*.js",
  "test": "mocha --timeout 30000 test/functional_api_crud.js",
  "pretest": "npm run lint",
  "posttest": "echo All tests have been run!"
}
```

When you first install eslint, you can run it in initialization mode and it will create a file based on the questions you answer. The command is:

```
./node_modules/.bin/eslint -init
```

I ran this and then made a few tweaks to the following two files:

```
// .eslintignore
node_modules
build
src
test
```

And the other file:

```
// .eslintrc.json
{
    "env": {
        "browser": true,
        "es6": true,
        "node": true
    },
    "extends": "eslint:recommended",
    "parserOptions": {
        "ecmaFeatures": {
            "experimentalObjectRestSpread": true,
```

## Chapter 13: Testing the NewsWatcher Rest API

```
            "jsx": true
        },
        "sourceType": "module"
    },
    "plugins": [
        "react"
    ],
    "rules": {
        "linebreak-style": [
            "error",
            "windows"
        ],
        "no-console": "off"
    }
}
```

It is as simple as running the following commands. The first just runs lint and the second runs lint and then the tests. Lint will list all the errors and you can keep fixing them until they are all gone. Here are the commands:

```
npm run lint
npm test
```

# Chapter 14: DevOps Service Layer Tips

It is a fabulous accomplishment to get the code all tested and deployed to production. Don't get too comfortable though, as it is quite another matter to manage the operations of a full-stack application. This chapter will present some key skills that will make your life easier when it comes to running a 24x7 operations for your service layer.

Chances are that you will experience some type of catastrophic failure before too long. First off, you absolutely want to do everything up front to put preventative measures in place. As the old saying goes "An ounce of prevention is worth a pound of cure."

You also need to put a plan in place for handling a crisis when it happens. You want to be in a position to have all of the information at your fingertips to make it possible to recover in the shortest amount of time. There are some general techniques that will be presented here, but you will have to come up with your own specific strategies that fit your own environment.

Let me make a brief comment about continuous integration and continuous delivery (CI/CD). If you are working on any kind of substantial project that will be going on for a while, or has multiple people contributing, you definitely need to implement CI/CD. Manually performing the tasks of building, testing, and deploying code gets old fast. Always remember that doing these things manually is prone to human error. If your DevOps process is not automated with full integration testing, then it is not really complete.

There is a wise saying that states "You have to slow down to speed up." You can interpret that to mean that a little investment up front pays huge dividends over and over. This ability to centrally coordinate CI/CD really helps groups with an agile process iterate more rapidly. Productivity goes up because of this automation and team downtime is reduced because integration bugs are not spread across the rest of the team. It is much more expensive in time and money to catch bugs in production, so there is a huge savings here.

## 14.1 Console Logging

Writing messages to a console output or log file is an age-old practice that might be useful if you can avoid being overwhelmed by too much logging and then be able to interpret the information to solve problems.

Node has the Console module that is useful to write trace output to. This is available in your application already, so you don't need to use a `require` statement. Here are some useful methods you can use that are found on the console object:

| | |
|---|---|
| `log()` `info()` `error()` `warn()` | Basic method for outputting text. It comes in several different forms all with the same function signature.<br><br>`console.log("we made it here %d", someVarNumber);` |
| `dir()` | It is useful to view an object you might have in your code. There are options available for this, for example to recurse further than the default depth of 2 levels.<br><br>`console.dir(someObject);` |
| `time()` and `timeEnd()` | To log elapsed time, you use these two methods. You will get the elapsed time when you do the following:<br><br>`console.time("start");`<br>`// some code operations that you want to time…`<br>`console.timeEnd("start");` |
| `trace()` | For showing a stack trace from the point in your code where this is called.<br><br>`console.trace("someLabel");` |
| `assert()` | This is the standard assertion usage commonly available.<br><br>`console.assert(valid, "Hi");` |

You can set up logging to go into a file that can be looked at. Sometimes logging will point you in the general direction and then you can use the debugger to further diagnose an issue. As another option you could send every log message to a special collection in MongoDB. You can set up MongoDB documents to have a time to live (TTL) before they are automatically deleted or set the collection as capped.

## 14.2 CPU Profiling

The V8 engine can provide you with CPU usage reports. The simplest way to do this is to launch your node process with the profile flag set, after running your test code, you stop the process and you will then have a file available to view. You would launch Node as follows on your local machine:

```
node --prof server.js
```

# PART II: The Service Layer (Node.js)

If you then run some test code, such as your load testing suite, you can get some idea of where your code might be spending most of its time. Once you have run your test code for a bit, you stop the node process and a file named something like "isolate-000001D213C4D490-v8.log" will be saved.

Next, open your browser and navigate to the Chrome V8 profiling log processor site at http://v8.googlecode.com/svn/trunk/tools/tick-processor.html. You can open your log file with this tool and see a nice, crisp layout of your code calls listed in order of where most of the time was spent. There is a command line version of the tick processor also available for download.

You will see reports that look as follows:

*Figure 80 - V8 Report sample*

*Figure 81 - V8 Report sample*

246

## Chapter 14: DevOps Service Layer Tips

You can also utilize the V8-profiler using a Node module to start it up in your code if you set up route handlers and want to profile things in production. The V8 engine, which is one of the components that Node.js is built on, lets you run an analysis of production running code. Profiling data can be sent to an external file that you can later open and view the results with. Here is the code to accomplish a profile run:

```
var fs = require('fs');
var profiler = require('v8-profiler');
profiler.startProfiling();

...do some processing...

var profileResult = profiler.stopProfiling();

profileResult.export()
  .pipe(fs.createWriteStream('profile.json'))
  .on('finish', function() {
    profileResult.delete();
  });
```

Here is a screenshot of the Chrome Dev Tool (press F12 while running Chrome) that is capable of reading the profile file.

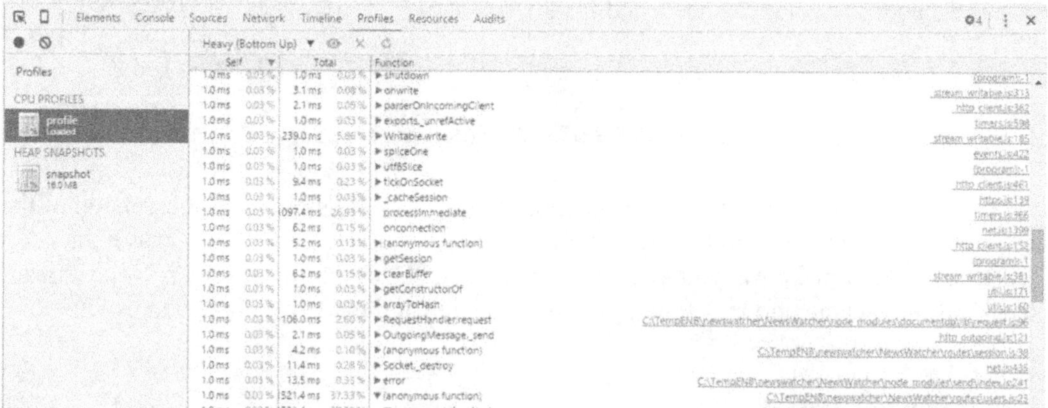

*Figure 82 - Chrome developer tool V8 profile report*

247

# PART II: The Service Layer (Node.js)

To read in your file, Click the **Profiles** tab, then click the **Profiles** pane and click **Load** to load your file(s).

With profiling turned on while a load test was run, I was able to produce a report that showed a few issues I could address. One of them turned out to be the use of `bcrypt` for password hashing. As seen in the profile analysis, you see a bottom-up view of the calling tree. If you expand until you see your functions, you can then understand where the call is being made from. Here, you see that the `bcrypt.hashSync()` function call took 19% of the time.

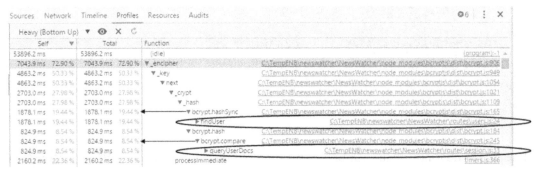

*Figure 83 - Profile drill-down*

Here is the code for what you see listed for the first issue:

```
findUserByEmail(req.db, req.body.email, function findUser(err, doc) {
    ...
        passwordHash: bcrypt.hashSync(req.body.password, 10),
```

It was obvious, given the name of the function, that I was not using the async version of the hash-generation function. Go figure, I should have known that `hashSync()` was not a good idea to call. As you know, you want to minimize CPU usage on the main Node thread. Checking the documentation, I found the async version and made the change to use it:

```
bcrypt.hash('bacon', 8, function(err, hash) {
});
```

**Note**: *It is helpful to give names to all anonymous functions, otherwise you will just see a lot of "anonymous" functions and it is harder to pinpoint what functions are in the profile listing.*

# 14.3 Memory Leak Detection

In a managed language framework, you don't directly allocate memory and subsequently free it up. You can do a `new` and `delete` of an object, but you still don't have control over that memory, such as the actual reclaiming of it, or doing things like having memory pointers into it. Instead, there is a garbage collector that keeps track of memory references and the GC decides when to run and when memory can be recycled.

The truth is, that with garbage collection running, you can still run into memory leaks that will eventually cause your application to either run slowly, to completely freeze, or crash.

You can obviously create a memory growth problem if you had something as simple as an array that you continually pushed data into and never free up. If you held on to an object reference permanently after you no longer needed it, the garbage collector will not ever reclaim it. A thorough code review of callbacks, closures, constructor functions, and arrays can be a starting point to finding memory leaks.

Ultimately, your brain might not be able to trace through all of the intricacies of your code and you will need to take memory snapshots that can be compared across time. Node.js applications are always built with several, if not dozens, of downloaded modules. You have to be suspicious of those, as well, as they might contain memory leaks.

You can watch the OS reporting of memory for your Node.js process over time and see what kind of graph you have and, if you see an ever-increasing amount of memory being taken up, you can then dive in and investigate. It is even not unheard of for people to resort to restarting their Node.js processes every day just to circumvent any memory leak problems. You might, in reality, not have a leak if, over time, you can see the garbage collector kick in and do its job. Compare memory snapshots over a 24-hour time span.

You can use the v8-profiler module that was previously used for CPU profiling to take memory snapshots. You can likewise have the output files viewed in the Chrome debugger.

```
var fs = require('fs');
var profiler = require('v8-profiler');
var snapshot = profiler.takeSnapshot();

snapshot.export()
  .pipe(fs.createWriteStream('snapshot.json'))
  .on('finish', snapshot.delete);
```

*Note*: Remember to never run your CPU profiling at the same time as you take heap snapshots. The overhead memory usage for the CPU profiler will inundate you.

## PART II: The Service Layer (Node.js)

I have set up a specific route through Express that can trigger a memory snapshot. That way I can insert this ability into my tests and have it available for me as part of a load testing run. Here is a screenshot of a memory snapshot file that is loaded into the Chrome browser debugger.

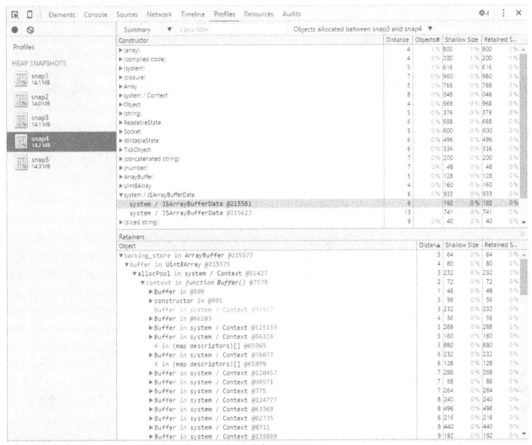

*Figure 84 - Chrome developer tool with memory snapshot*

The Distance column shows you how many steps removed from the root object the memory reference is. You can usually assume that the object with the shortest distance is the one causing a memory leak. The Shallow Size is just what this one usage is taking. Retained Size gives you all of the space that would be freed up that this object is referencing and thus holding on to. This is only true if those objects are also no longer referenced by anything else.

There are different views you can try out, such as Containment, which helps you also view low-level memory internals.

## 14.4 CI/CD

Do you work on a project where the tasks of building, testing and deploying code are all done manually? Why is that? From my experience, the reason teams don't automate manual tasks is typically because they either lack incentive, or they lack the knowledge. Lacking incentive is really a poor excuse. There is plenty of incentive if you honestly look at the return on investment of implementing a CI/CD process. If your DevOps process is not automated with full integration testing, then it is not really complete.

I once heard a speaker at a conference say to developers – "It ain't done until it is automated". He was saying that you need to automate everything from the check-in, testing and integration through building and deploying. In this book, I will say the 'D' in CD stand for delivery. This means that code is served up and ready to be deployed. We will, however, take the process all the way through to delivery and can say we implemented a CI/CD+ solution.

Another statement I have heard and that I often repeat is that of – "You have to slow down to speed up." You can interpret that to mean that a little investment up front pays huge dividends over and over. This ability to centrally coordinate CI/CD really helps the agile environment iterate more rapidly. Productivity goes up because of the automation and team downtime is eliminated because integration bugs are not spread to the rest of the team. It is much more expensive in time and money to catch bugs in production, so there is a huge savings to be had here.

Now that you have the incentive, let's assume that it really comes down to knowledge. After reading this, you will be informed and will no longer have any more excuses preventing you from implementing CI/CD.

## Many tools and ways to accomplish CI/CD

Implementing CI/CD can, of course, be a complex undertaking. No one article or book can tell you everything you need to know because there are so many tools out there. No one solution will work for all projects. All projects are unique in their code structure and frameworks used. We will only scratch the surface and only do so for a Node.js project in a very narrow niche. We also narrow our focus down by relying on a PaaS solution.

If you choose not to go the route of setting up a complete CI/CD tool, you can still automate all of this on your own and launch things manually. For example, you can implement a series of Gulp tasks that would do a lot of what a CI/CD system would do.

*Note: You can do everything in your DevOps environment through the command line. Once again, I have chosen to take the route of using UIs where available.*

PART II: The Service Layer (Node.js)

# You must have developer tested code

As a preparatory step, a developer will have prepared some code and privately tested any new feature before submitting it to the CI/CD process. The testing should be as thorough as possible in the context of the rest of the integrated system. If that is not possible, then it can be tested as an isolated unit, perhaps with stubbed out or mocked functionality injected. The testing should be runnable by a developer by typing "npm test", which will run lint and then all tests for that part of the code.

Once the developer has had their own private verification completed, they can do their code check-in that will then be staged to go into the pipeline for consideration. If you are using Git and GitHub, the code would be pushed to the branch you designate for the CI/CD to be triggered from. You would not necessarily push it into the master branch. The master branch should only be used for the code that is running in production. You could push to master and tag that but not have it deployed to production, but to test and then have a step that requires manual acknowledgment to push that to production.

CI/CD is all about increasing your iteration speed and the quality of everything written. Of course, you have to provide high-quality comprehensive test suites to achieve this. Once your code is ready to commit, then the pipeline workflow of CI/CD takes place. So what does the cycle look like? Let's go through each of the steps. Here is a simple diagram to show you the pieces that would make up a simple Node.js CI/CD system:

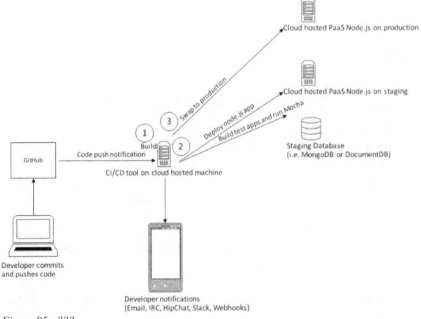

*Figure 85 - ???*

252

## AWS tools

There are a lot of companies making tools to help with doing CI/CD. I have loved using the open source tool Jenkins in the past and it is a good choice, and there are even companies hosting it in a PaaS environment if you don't want to install and manage it yourself.

This book will use AWS CodePipeline as the overall orchestration tool to mage the build, test and deployment steps. The build and test are done with AWS CodeBuild. Deployment can be done with CodeDeploy to deploy to ECS for a Microservices architecture. In our case, we use Elastic Beanstalk. There is a direct integration to that in CodePipeline, so CodeDeploy will not be used.

## Step One: Install

Once the code is committed into the source code repository then the code needs to be staged in the overall application. CI/CD tools actually are able to listen to GitHub and will kick off immediately upon seeing a code push. The idea is that a CI/CD system can see the push and do a local clone of the branch so you have a freshly integrated code version to now work from in the CI/CD system.

All code dependencies such as NPM modules would need to be updated in this clone and that is part of what the CI/CD would kick off. This is done with an npm install command. Since we are wanting to get a production environment up and running, we will have to do an extra npm install to get the DevDependencies since those don't get installed if the environment variable NODE_ENV is equal to "production". This will be explained again as we go through the steps.

## Step Two: Build

Once all the dependencies are installed, we can do any build and test steps. In our case, the Node.js code is set to go, but there would be a build step required for our React code. Then things like Lint, can be run. If Lint passes, then the Mocha tests can be run.

Comprehensive testing is central to getting CI/CD working. This step could run a thorough suite of tests at all levels, such as Unit, integration, feature, load, performance and UI automation testing. A perfect pass is expected and you would get generated reports to show that all went well and also produce code coverage reporting and have memory leak detection.

## Step Three: Post-Build and deploy

This is the final step that can generate the files necessary to deploy and actually do the deployment if you like. Since we had development dependencies like Lint and Mocha installed, we can back those out and just have dependencies needed for the production environment. The deployment can be automatic, or an email can be sent to a person to do manual acknowledgment to approve it. If your team is in favor of TIP (Testing in Production) you could deploy to a reserved hidden portion of your production environment.

You should have a testing environment that is as close to production as possible. One thing to remember is to also have in place a dedicated database for testing purposes hosted in PaaS that is always available. You do not need to continually deploy there if it is simply the location of the data and nothing needs to be preconfigured. You can run some cleanup script as part of the CI/CD step here if that is necessary and also some pre-population script if you require certain Documents to be in place. PaaS offerings for MongoDB and others are available as document-based databases.

If the testing fails, you would be notified and the CI/CD process would not proceed any further. You would want failures automatically entered into an issue tracking system to officially track and resolve issues.

## NewsWatcher CI/CD with AWS CodePipeline

Here is a quick introduction to how to set up a CI/CD process with AWS CodePipeline.

Make sure you are already signed into GitHub, and then in the AWS console select **CodePipeline** from the **Services** drop down. Here is a brief listing of some of the steps. I have left out some of the details as they should be obvious.

1. Click **Get Started.**
2. Name your pipeline and click **Next step.**
3. For "Source provider" select GitHub and click **Connect to GitHub.**
4. Click **Authorize aws-codesuite** and confirm your password.
5. Select the **Repository** and **Branch**. I selected the master branch. Click **Next step**.
6. Select **AWS CodeBuild** for the Build provider.
7. Select the runtime to be Node.js and the latest version.
8. For the Build specification, select to use a buildspec.yml file.

Here is what I have in the buildspec.yml file:

```
phases:
  install:
    commands:
      - echo Installing Node Modules...
      - npm install
      - npm install --only=dev
  build:
    commands:
      - echo Build started on `date`
      - echo Building React Web application
```

## Chapter 14: DevOps Service Layer Tips

```
      - npm run build-react
      - echo Performing Test
      - npm test
  post_build:
    commands:
      - echo Final build, without devDependencies
      - rm -rf node_modules/
      - npm install
artifacts:
  files:
    - '**/*'
```

Once you are ready to go, you can merge and push a code change up to GitHub and watch in the AWS console with CodePipeline as each step progresses and finally see the code pushed out to production. Here is a screenshot showing the full process after it is run. If you happen to have an error in the CodeBuild step, then you can open that up and see the details and see how to fix the error.

*Figure 86 – AWS CodePipeline usage*

255

# 14.5 Monitoring and Alerting

The AWS Elastic Beanstalk management portal has a Monitoring page with which you can view key machine performance metrics. You definitely want to open the portal and look at what is available to be monitored. Not only can you look at trending charts, but you can also set up alerts that send you emails, or even get text messages in the case of thresholds being crossed. Here is the Monitoring page:

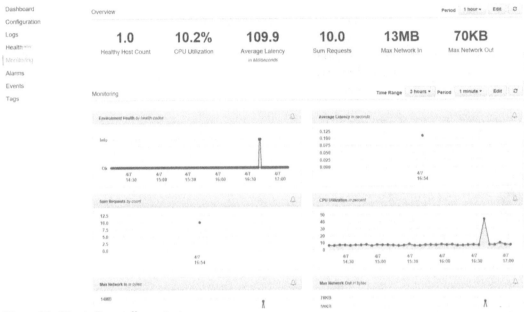

*Figure 87 - Elastic Beanstalk monitoring page*

You can click **Edit** and select what you want to be graphed out. You can also click the alarm bell icon and create an alarm that would notify you if you crossed some threshold.

Alternatively, you can open up the CloudWatch management console in AWS and explore similar capabilities for metric viewing, logs, events and alarms.

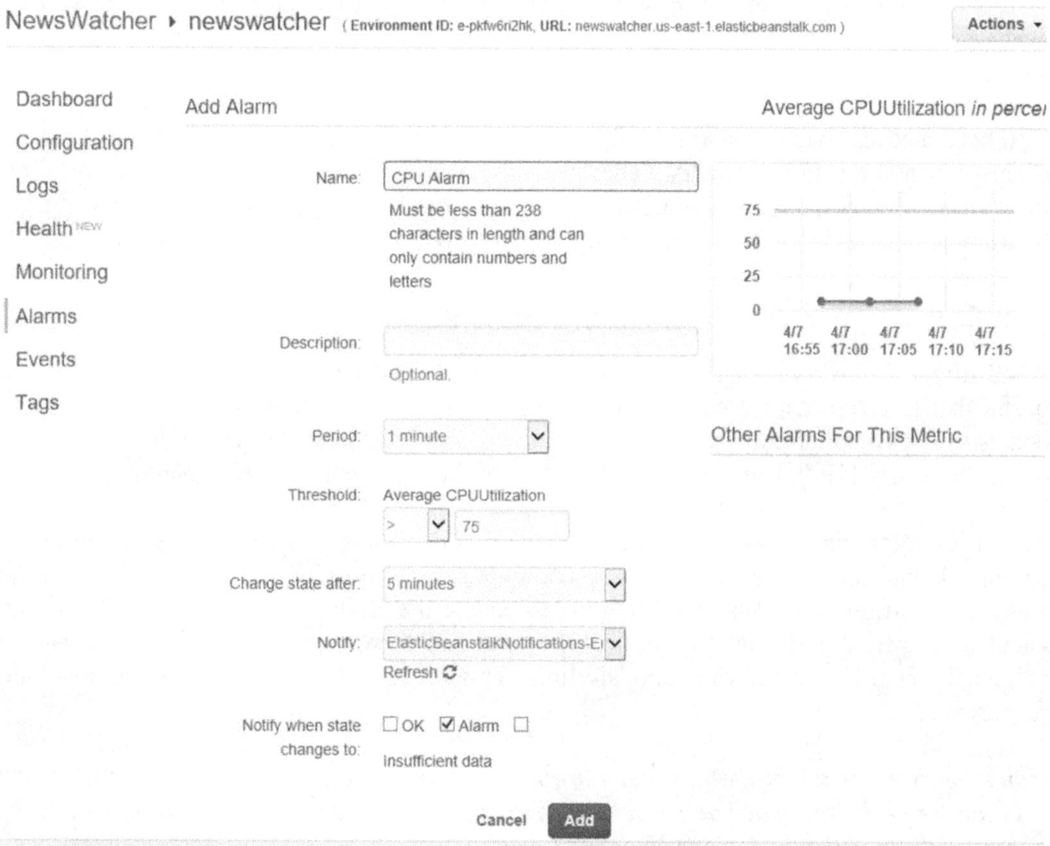

Chapter 14: DevOps Service Layer Tips

*Figure 88 - AWS CloudWatch management console*

One thing to know is that the Node.js process can crash and if that happens, it needs to be restarted. In an IaaS environment, you would be responsible to have some mechanism to detect this and restart your Node.js process.

There are NPM downloads such as "forever" and PM2 that do this. With AWS Elastic Beanstalk, it uses Nginx and the restarting is handled automatically for you. This is another example of how PaaS really done right can make your life easier.

If your application is not available, reliable, and performant, then your customers and your business will suffer. The goal of monitoring and alerting is to maintain application availability, reliability, and performance. You do this by implementing Application Performance Monitoring (APM). This allows you to discover problems before anyone else does and then to achieve resolutions in the least amount of time.

PART II: The Service Layer (Node.js)

To begin implementing an APM strategy, you need to instrument your code and surface events, logs, and metrics. Then you use an APM tool to chart out performance metrics and set up alarms.

You need metrics so that you are not flying blind. A pilot can fly a plane in the dark because of instrumentation and telemetry. The last thing you want to have to do is remote into individual servers and start poking around to search for a "needle in a haystack". Instrumentation and monitoring is the only way you will be able to scale and survive.

## AWS X-Ray for code instrumentation APM

AWS offers X-Ray as a way to instrument your code and have the telemetry viewed in a portal that lets you inspect each of the traces that your application is processing. A trace consists of information for what is occurring in your service layer. For example, you can see a trace for every HTTP route endpoint and each of the verbs that are being handled.

To get X-Ray tracing to work with your Elastic Beanstalk application you need to turn on a setting. Go into the Elastic Beanstalk console and then into the Configuration settings for the Software Configuration and check the box to enable the X-Ray daemon. This will run the agent on the EC2 machines that collect the data and forward it to the location it can be collected together for viewing and alerting. This will enable XRay to get at machine resources.

*Note: NewRelic is a tool that has been around for some time and is a very comprehensive solution for gathering machine resource usage and transaction instrumentation. Like AWS X-Ray, you can add code to your Node.js project to instrument the sending of telemetry. You can customize what is sent back and add custom metrics and events. You can also add a script to your HTML client to instrument from the client side all usage. NewRelic has a rich capability to set thresholds and alert on them.*

There are costs associated with using X-Ray, but if you have to process a lot of tracing in order to incur any cost. Please look at the AWS documentation for an explanation of the cost. You can control what it actually captures on a route-by-route basis. You can completely ignore a route, or set it up to only have a certain percentage of traffic captured.

X-Ray understands how to collect tracing information for transactions that are handled by your Express usage in Node through the aws-xray-sdk npm module.

*Note: To download the module, you need to be running your machine console as admin. Then you can run "npm install aws-xray-sdk –save".*

## Chapter 14: DevOps Service Layer Tips

Here is code to use AWS X-Ray in Node JavaScript:

```
var AWSXRay = require('aws-xray-sdk');
app.use(AWSXRay.express.openSegment('NewsWatcher'));

// All your API routes
app.use('/api/users', users);
app.use('/api/sessions', session);

app.use(AWSXRay.express.closeSegment());
```

If you then go to X-Ray in the AWS console you can see the traces. You can also see all of the AWS resources in your service map. Here is what the trace view looks like. You can click on a trace and see its details.

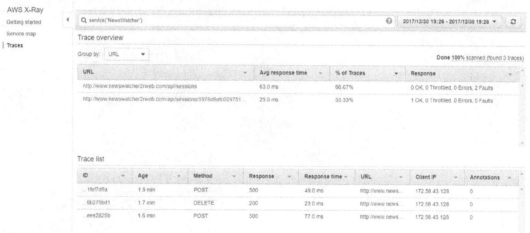

*Figure 89 – X-Ray trace*

If your code has an unhandled exception, then you could actually see the stack trace in the details of the trace. You can also add extra information to any trace as follows:

```
var segment = AWSXRay.getSegment();
segment.addMetadata("errorHandler", err.toString());
```

If you were using the AWS SDK in your code, you then you can add some code to make it aware of those calls, and that usage will be instrumented. As of yet, there is no support for MongoDB, but I anticipate that there will be soon.

It is possible to put in code to measure any part of a transaction trace such as some processing code you want to have a further breakdown of to see its timing in the overall trace time. This can be synchronous, or asynchronous code.

259

# PART II: The Service Layer (Node.js)

The idea is that there is an incoming request to your Node.js service, such as a POST to an API endpoint. A single trace ID is assigned to that request and no matter what happens up to the very end of that request, every subpart of work that goes on shares that ID. That means all of the work in your code, calls to other backend HTTP services. Any use of other AWS resources is all tied together and viewable under that single ID in the X-Ray console.

# PART III: The Presentation Layer (React/HTML)

Part three of this book will teach you about the presentation layer of a three-tier architecture. In doing so, the sample application will be extended to bring it to a state where the UI is functional and interacting with the services layer. Before reading this chapter, you should obtain a basic understanding of HTML.

In this part of the book, I present the technology of React as a framework that fits nicely into the overall architecture as a way to bind data to and from the service layer web service into a UI. The UI will be rendered as an HTML Website and also as a Native mobile application.

If you have followed along in the previous chapters of this book, you will realize that there is a service layer built and tested for the UI to connect to. This is very important, as the Node.js application will perform the dual role of servicing not only the HTTP/Rest API, but also of serving up your React HTML and JavaScript files etc. to a browser as a SPA application.

*Note: It is not the intention of this book to be a comprehensive guide to UX design or SPA web design. I will only touch on some concepts and then stick to a narrow technology presentation. There are whole books each devoted to React and React Native, so you can understand that we can only cover the basics of how each work to get you going on this topic.*

# Chapter 15: Fundamentals

I will now go over the fundamental concepts of the top tier in a three-tier architecture. You will see what capabilities are essential and find a list of questions to consider when doing your code design. You can then get into the specifics of React as a JavaScript UI framework and learn how it will be used with the NewsWatcher sample application. One of the reasons that React was chosen as a technology is because it fulfills the needs of this top layer of the application architecture and does so with the JavaScript language.

## 15.1 Definition of the Presentation Layer

Any application that requires user interaction will need a presentation layer. Humans need a presentation layer to view data and to allow them to input data. For example, an online bookstore would be presenting a list of books to a user to browse through. Each book offered, might contain a photo of the book and the data associated with it such as the title, author, description, publication date, and cost. Input gathered from the user would be things like book orders, reviews, and customer service questions.

### MV* and SPA designs

You need to employ the techniques of abstraction and componentization in all the layers of an architecture. This is no different with the coding in the presentation layer.

One of the benefits of choosing a framework for your presentation layer is that most frameworks are set up to employ some type of MV* pattern that lends itself to an organized set of components that make up your code.

Make yourselves acquainted with MV* design patterns and with what a Single Page Application (SPA) design entails. There are several excellent books available and plenty of online material to study. There are many things about SPA designs that make them a great choice today.

*Note: You need to be aware that if you are concerned about SEO (Search Engine Optimization) you might need to switch some of your page rendering to be server-side and not be fully rendered as a client-side SPA. This is sometimes required for search engines to index your site. React can be set up to give you the balance you need for this. There is a short chapter devoted to how to do this.*

PART III: The Presentation Layer (React/HTML)

# Presentation layer planning

There are many decisions that go into creating a presentation layer. The very first step involves planning for what type of interactions your users would want to take. You will want to make an initial sketch of the UI of your application early on before coding anything up. This would happen concurrently as you design the service layer. It is wise to pursue a simultaneous bottom-up and top-down approach.

Knowing what operations and workflows are needed is another first step in fleshing out a presentation layer. The following questions are used to help you determine the design of a presentation layer:

- Have you done any of the following? sketching, prototyping, storyboards, surveys, contextual inquiry, stakeholder interviews, A/B testing, wireframes, sitemaps, personas, scenarios. What about human interaction, usability, and accessibility studies?
- What are your data security and privacy requirements?
- How do people sign in and become authorized?
- Are there multiple steps that are progressively revealed, one after another?
- Is there a need for a customizable UI?
- What are your globalization and localization requirements?
- What devices are you targeting, such as desktop and mobile platforms?
- How can you keep data presentation to a minimum to not overwhelm the user?
- What form is data best presented in?
- What is the business need that can be accomplished?
- What accessibility requirements are there?
- What are the navigation levels of the UI? Can you map out how the navigation works?
- Do you need a user feedback mechanism?
- Do you have offline requirements?
- How is state stored in the client? Centralized across all view hierarchy?
- How will the UI be deployed and updated?
- How will users enter data and what tests are needed to validate it?
- Can you map out the multi-step data entry forms and show the branching conditions?

The answers to these questions should be carefully documented. Before you roll anything out into your production environment, have experts reviewing everything.

React is a great choice as a UI framework and fulfills all the needs of the presentation layer of the application architecture. It uses JavaScript, so it fits in perfectly with the overall development stack we have been pursuing. You can now learn the specifics of React and see a practical implementation of how it is used with the NewsWatcher sample application.

# 15.2 Introducing React

React is a framework that gives you the ability to dynamically render HTML content in a browser web page. You write JavaScript to go alongside your HTML markup. The JavaScript code can leverage other available libraries to do things like HTTP/Rest requests, data binding and navigation and much more. The overall capabilities of React provide the mechanisms to enable you to build a SPA, a traditional server-side rendered web site, or a hybrid.

There have been other similar frameworks that have appeared at the same time as React such as Knockout.js, Ember.js, Angular, Vue.js and Backbone.js. React is my choice in this book as it has many things going for it, such as being widely adopted, and also because it is officially sponsored by a large company (Facebook) as an open-source project. You can be up and running on the desktop, mobile web, and even produce native mobile applications in a short amount of time.

*Note:* *You see me referring to React as a framework. I do so in the larger sense. React, along with your chosen additions for other functionality, provide a framework that you built your application presentation layer with. That is the point of a framework, it is extensible. It just so happens that React is not as prescriptive and feature capable as something like Angular. It is up to you to combine it with the other libraries of your choice to give you all of the features you need.*

React frees you from some of the laborious code you used to have to write to do DOM (Domain Object Model) manipulation. Because of its ability to bind and affect DOM elements, you no longer need to utilize libraries such as jQuery.

React makes HTML dynamic so that data values flow back and forth for you. To use React, you write component files that render to the DOM. Be aware, however, that React itself does not provide control elements or styling for you, as that is left to the HTML markup and other libraries such as material design, Bootstrap and many more that are available.

The React library is consumed by placing a script inside your HTML file that pulls in React as JavaScript to execute from a CDN. You can also locally bundled React through Webpack and then pull it in that way.

There are third-party tools that act as UI design studios, where you can drag and drop elements and play around with them and have React code generated and edited. This is handy to be able to visualize a component in an isolated way so as to save you time. Otherwise, you need to see the component in an overall application. Look into tools such as Storybook and React Styleguidist.

PART III: The Presentation Layer (React/HTML)

*Note*: *You might recall that the Node.js Express module has the concept of serving up templates of "HTML-like" files and binding data to them on the server-side, so that they arrive on the client side all filled out. Express supports many template formats such as Jade, EJS, mustache, and handlebars. I don't recommend this server-side data binding technique. Instead, I give preference to simply serving the HTML that works as a SPA application, with the client side using React to request data through a Rest API with JSON and then doing the binding on the client side. This alleviates the back-and-forth HTML page requests. With a SPA, all of the navigation and page rendering is done on the client side, not on the server side. This is similar to what you would need to do if you were to develop a native mobile application. You can also stick with HTML instead of having to learn a new template markup syntax such as Jade.*

## 15.3 React with only an HTML file

You could pull in the React library with a script markup in an HTML file. This is not a good idea, but I show it as a way to understand what ultimately has to happen to bring any code into an HTML page. React is just a library like anything else that runs as a script in a browser that exists in an HTML rendered page. You can place this text in a file and open it with your browser and see that it works.

```
<!DOCTYPE html>
<html>
<head>
  <meta charset="UTF-8" />
  <title>Hello World</title>
  <script
    src="https://cdnjs.cloudflare.com/ajax/libs/react/15.4.2/react.js">
  </script>
  <script
    src="https://cdnjs.cloudflare.com/ajax/libs/react/15.4.2/react-dom.js">
  </script>
  <script
    src="https://cdnjs.cloudflare.com/ajax/libs/babel-standalone/6.21.1/babel.min.js">
  </script>
</head>
<body>
  <div id="root"></div>
  <script type="text/babel">
      ReactDOM.render(
        <h1>Hello, world!</h1>,
        document.getElementById('root')
      );
  </script>
</body>
</html>
```

What you want to do, is install the necessary tooling on your developer machine that can be used to do a build of your code to produce a bundled file that has everything in it from your project to use. This means React is bundled with your web site upload of static files. You end up with a generated HTML file that looks something like the following:

```html
<!DOCTYPE html>
<html>
<head>
  <meta charset="UTF-8" />
  <title>Hello World</title>
</head>
  <body>
    <div id="root"></div>
    <script type="text/javascript"
      src="/static/js/main.35d639b7.js"></script>
  </body>
</html>
```

The JavaScript file that you see being pulled in with the script tag is generated in your build process on your machine and contains all the needed React code and your own code that is necessary to load and run your SPA page. The next section will walk you through how to arrive at the file shown above in the simplest way possible.

## 15.4 Installation and App Creation

If you really want to go through all of the work of installing each of the modules and tools for doing react development you certainly can do that. To do so, you can go to the official React website and follow their downloading instructions. You can start a project on your own and do npm installs of modules such as react and react-dom. Then you would need to install tools for cross-compiling and bundling. You would want to get tools installed from NPM such as webpack, Babel, Autoprefixer, ESLint, Jest, and other tools. You would need to become very familiar with the following terms and how to download and manage tools for each.

**cross-compiling:** Allows you to use the latest JavaScript syntax and have it turned into JavaScript that can run in older browsers.

**bundling:** This is how to gather all the files together for much better performance and a controllable loading capability. Otherwise, you have to reference the minified JavaScript files in your HTML and take the hit on the initial page load. Going with bundling also gets you prepared for being able to deploy as a native application on a mobile device, as that is the only way to accomplish that.

PART III: The Presentation Layer (React/HTML)

Instead of installing the tooling on your own, you can simply install a tool that populates a React project for you with some example code and all of the tooling.

## Using create-react-app

The task of getting the initial project files in place to have the traditional "Hello world" application running may seem a little daunting at first. The best approach is to install and use the `create-react-app` utility. Install this tool as follows:

```
npm install -g create-react-app
```

You could now run this and you would have an application all ready to run. In our case, we already have an existing folder with a Node.js application in it. What we will do is create the React application in a separate folder. Then simply copy over the necessary folders from the created project over into the Node project. This is the best approach to take for now. Run the following on the command line in a completely new directory. This will create your basic React application that you can pull from:

```
create-react-app my-app
```

Copy over the "public" and "src" directories. Look in the package.json and the .gitignore files and bring over what is missing in the ones in the Node.js files. While you are editing the .gitignore file, add ".chrome" as a line, as this will be needed later. For the package.json file that meant just bringing over a few of the script commands. I renamed them, so as not to conflict with anything I had for my Node.js project.

```
"start-react": "react-scripts start",
"build-react": "react-scripts build",
"test-react": "react-scripts test --env=jsdom"
```

You also need to copy the Node.js package.json lines that were for the following dependencies: react, react-dom and react-scripts. Then you can run npm install and have those ready to use.

To run the React application, you can now run the command "react-scripts start" and that would be served up for you on your local machine. We have put a script command in package.json for these. You can thus run "npm run start-react". If you run that, you will see the following launched in the Chrome browser:

## Chapter 15: Fundamentals

*Figure 90 - Starter project screen for React install*

***Note***: *Node.js has its own commands to get an initial application structure set up and configured on your development machine. It can be confusing on what to do for your initial project creation, as you have just seen that there is also a command line utility to create a React application. There is a lot more to a React install, however, especially if you use a framework setup, like the create-react-app utility. Thus I had to essentially get each project set up independently and then bring one over to the other. There are other utilities out there such as Yeoman that can give you the complete scaffolding for everything in one single initialization. There is also a generator named react-fullstack that can be used as a starting point for your project. Of course, you can also clone anyone's GitHub application, such as mine, and have everything all ready to use.*

## Serving up React from Node.js and Express

Before getting into the heavy coding in React, make sure that the initial React page is served up correctly from the Node.js service. The details of making your Node.js Express application serve up the React index.html and JavaScript code as a Web Page are extremely simple. Here are the bare minimum lines of code you need to do that.

```
const express = require('express');
const path = require('path');
const app = express();

app.get('/', function (req, res) {
  res.sendFile(path.join(__dirname, 'build', 'index.html'));
});

app.use(express.static(path.join(__dirname, 'build')));

app.listen(3000);
```

PART III: The Presentation Layer (React/HTML)

In the next chapter, you will see the details of the NewsWatcher application code for the presentation layer. You will see how these lines are incorporated into the server.js file. To run it with Node, you put in those lines shown above and then need to first do a build and bundle of the React code before the Node service can be started up to serve it. To do the build and bundle of React, you run the following:

```
npm run build-react
```

You could then press F5 in Visual Studio Code and then your application is once again being served up, this time through Node, and you can go to http://localhost:3000/ in your browser to see it. If you are curious, you might want to inspect the build directory that is created and look at each of the files to see exactly what is being served and run on the client browser.

## The index.html template and generated one

If you recall, I showed some HTML that was all self-contained. It had everything needed to run a React backed web page. It had the scripts necessary to use the React libraries. However, with the project structure created with the create-react-app utility, it works completely differently. You still have the index.html file. That is used as the starting point for the one being generated. You will notice that it does not have any scripts pulled in. It looks as follows:

```
<!doctype html>
<html lang="en">
  <head>
    <meta charset="utf-8">
    <meta name="viewport" content="width=device-width, initial-scale=1">
    <link rel="shortcut icon" href="%PUBLIC_URL%/favicon.ico">
    <title>React App</title>
  </head>
  <body>
    <div id="root"></div>
  </body>
</html>
```

This HTML file is serving as a point template file and has that same div with an id of "root". By template, I mean, it is taken at build time and altered and the final index.html file is created and placed in the build directory. Here is what that looks like:

```
<!DOCTYPE html>
<html lang="en">
<head>
  <meta charset="utf-8">
  <meta name="viewport" content="width=device-width,initial-scale=1,shrink-to-fit=no">
```

```html
    <meta name="theme-color" content="#000000">
    <link rel="manifest" href="/manifest.json">
    <link rel="shortcut icon" href="/favicon.ico">
    <title>React App</title>
    <link href="/static/css/main.c17080f1.css" rel="stylesheet">
</head>

<body>
  <div id="root"></div>
  <script type="text/javascript"
    src="/static/js/main.35d639b7.js"></script>
</body>

</html>
```

## The starting point for your code

We can finally get to the real point you have been waiting for. You can now learn what the starting point is for where you place your own custom code for your React application. This code belongs in the index.js file found in the src folder. This is what is used at build time to be the starting point of your React app. Here is the file that the React build tools know about and use to bootstrap your application:

```js
//index.js
import React from 'react';
import ReactDOM from 'react-dom';
import App from './App';
import './index.css';

ReactDOM.render(
  <App />,
  document.getElementById('root')
);
```

You see the same `ReactDOM.render()` call that was in the initial self-contained HTML file. It is just split out and used at build time to be pulled in. What you see here is that the index.js file is importing the App.js file. In there you will find the definition of the component named App that is being rendered. React code is based on the concept of components that are assembled to build the UI. Eventually, it all leads down to HTML.

The index.js file is used by React to do the initial element insertion into the div that had the "root" id. This code uses the App.js file as the starting point for where you place all of the JavaScript and HTML tags to do the rendering of what your UI will be. Here is what the initial contents are for the rendering as found in the App.js file when we created it with the `create-react-app` command:

```
// App.js
import React, { Component } from 'react';
import logo from './logo.svg';
import './App.css';

class App extends Component {
  render() {
    return (
      <div className="App">
        <header className="App-header">
          <img src={logo} className="App-logo" alt="logo" />
          <h1 className="App-title">Welcome to React</h1>
        </header>
        <p className="App-intro">
          To get started, edit <code>src/App.js</code> and save to reload.
        </p>
      </div>
    );
  }
}
export default App;
```

When you run "`npm run build-react`" from the command line, this builds and bundles everything for usage to be served up from your Node.js service. After the build command completes, you will find that you have a new folder named "build". It is in here that the files are placed that are fully capable of being used as pages to be loaded and seen in a browser. For example, you will now find a new index.html file in the build folder that was shown already.

The big difference now between this file and the one in your public folder is there is a style sheet and script file brought in that are used to run the React application. This is why a Node.js server can be set up to serve up the build folder index.html file and also how it can get at its own built files for all the React and provided JavaScript files to run.

## Debugging React code

You will want to know how to debug this new React code. Debugging the Node.js code was simple. You just placed a breakpoint in your Node JavaScript code and you stop on it when that is executed and step through it. Section 18.3 will contain more on this topic.

For React code, it is really easy to debug your code in the Chrome browser. You can get to the code and set breakpoints by selecting from the Chrome menu **More tools->Developer tools**. You can then be in an environment where you can also inspect the DOM and look at console output and much more.

# Chapter 15: Fundamentals

You will need to run Node separately now, either from the command line ("npm start"), or in VS Code you can press F5. Then you can go to http://localhost:3000/ in Chrome and with the developer tools, debug the React code. You can, at the same time, debug server-side Node code in VS Code as it is running.

You can look up online to find out how to use the Chrome Developer tools. To set a breakpoint, you just open the **Sources** tab and then on the left side you will find the folders with the code in it. Go to the static folder and in there, will be the js folder. Open the views folder and select a file such as loginview.js and you can set breakpoints and step through the code.

The nice thing that is happening, is that you are looking at your code before it was cross-compiled and bundled. There are mapping files created behind the scenes that allow the actual code that is being run to be mapped to the original lines for your ease in debugging it all. Here is what it looks like with NewWatcher running and hitting a breakpoint in the code.

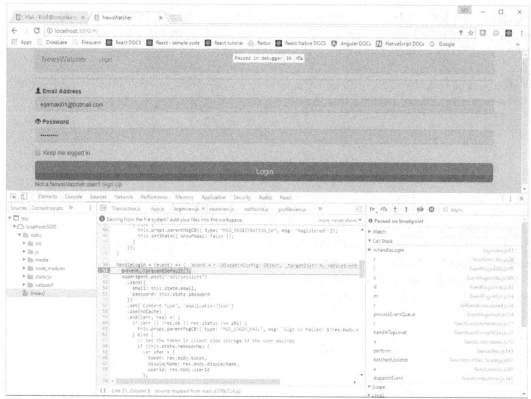

*Figure 91 - Chrome Developer tools*

273

PART III: The Presentation Layer (React/HTML)

# 15.5 The Basics of React rendering with Components

Perhaps you have heard that React does not use HTML files like a traditional Web Server might serve up. React instead uses code files to create components that get pieced together to create your UI. It is all done through standard DOM manipulation code at the lowest level. Don't be thinking you are getting away from HTML markup and CSS styling by any means. There will ultimately be just as much of that required to create your application.

With all the React component abstractions you piece together, at the lowest level you will end up providing HTML markup that gets rendered to the DOM. You have all the standard set of HTML elements to pull from. You will use tags such as `<p>`, `<a>`, `<ul>`, `<button>`, `<input>`, `<iframe>`, `<img>` etc. You can also add content, attributes and styling.

React has you provide HTML, but you do it in a JavaScript file (.js or .jsx) instead. The code is JavaScript that allows certain additional syntax. These files are cross-compiled into the actual JavaScript code that a Browser can run.

At the start of this chapter, you learned that there is a top-level index.html file that has a div element in it as follows (see the line in bold type):

```
<body>
    <div id="root"></div>
    <script type="text/javascript"
src="/static/js/main.3dd1fbc9.js"></script>
</body>
```

If you understand a little about browser DOM manipulation APIs, you know that you can create anything you want in the DOM through code. For example, you might insert a `text` element using JavaScript code. Furthermore, you can reference any existing element by an ID you give it. In that way, you can find it in code and alter it. This means you could find a `div` and add a child HTML element to it. That child element would then show up in the rendered UI as well.

Here is some plain HTML code with no React usage at all. You can place this in a file and open it in your browser to try it out. It uses some JavaScript code inside a script tag. The standard `document` object available in browser JavaScript is used to access your web page. It is part of an agreed-upon API standard available in any browser. The code creates a `text` element as a child of the `div` element.

## Chapter 15: Fundamentals

```html
<!DOCTYPE html>
<body>
<div id="root"></div>
</body>
<script>
  var textnode = document.createTextNode("Welcome to React");
  document.getElementById("root").appendChild(textnode);
</script>
</html>
```

React employs this exact mechanism to take anything you want to be rendered and presented on the page. To begin with, React is rendering to the main div you saw above with the id of "root". There is a starting piece of code in a file named index.js that is used as the page gets rendered. At the time the initial React application was created, it looked as follows:

```js
// index.js
import React from 'react';
import ReactDOM from 'react-dom';
import App from './App';
import './index.css';
ReactDOM.render(<App />, document.getElementById('root'));
```

This code eventually gets built and morphed into another JavaScript file that utilizes the React API to make calls that use the `document` object and do things like call appendChild. This is the first piece of code that the React library uses to render something.

The call you see to `ReactDOM.render(...)` is accepting as the first argument what element you want to render. App is a component from another file that was imported. The second argument to the `render()` call is telling React where to render the UI. That is where we are referring to the "root" `div` in the index.html file.

The last file to look at in the process of understanding how React does DOM rendering, is the file that contains our custom defined component. In this case, the file is named App.js. This then is the pattern for the code that you will create over and over as you define the UI you want to be rendered. Here are the contents of that file:

```js
// App.js
import React, { Component } from 'react';
import logo from './logo.svg';
import './App.css';

class App extends Component {
  render() {
    return (
      <div className="App">
```

275

```
        <div className="App-header">
          <img src={logo} className="App-logo" alt="logo" />
          <h2>Welcome to React</h2>
        </div>
        <p>edit <code>src/App.js</code>to reload.</p>
      </div>
    );
  }
}
export default App;
```

React is a JavaScript library that exposes an API. Part of that API is a class you can derive from that it understands. Component is a class you derive from and write code to contain your own UI to render. App is your highest parent control and is derived from Component. You will create a complete hierarchy of other components you define and make use of. In the app.js file, you export the App component so that it can be imported and used in the index.js file.

Notice the first line of app.js that does an import of React. This must present even though you don't see the usage of that object. This is because the JavaScript code is transformed by Babel, and will be used in that code.

*Note*: By convention, you start your custom components with an uppercase letter. This then helps you distinguish these from HTML elements that are always all lower case (i.e. div).

# The render Method

The Component class requires you to provide a method named "render". React takes your App component and calls the `render()` method to get anything that needs to be rendered. Be aware that this is not actually a JavaScript file that can be run. As explained, it is used in the React build process and turned it into something that can be run. You may have seen that the return in the render method is not really returning anything acceptable in JavaScript.

This use of JavaScript is what is referred to as the JSX syntax. It allows you to mix HTML tags and JavaScript right in one file. For example, you can have conditional code that determines what elements should be rendered on a page. Look at the following example:

```
render() {
  if (user.account.total > 1000) {
    return (<h1>You maintained the minimum required balance!</h1>);
  }
  return (<h1>You are below your minimum required balance</h1>);
}
```

## Chapter 15: Fundamentals

Be aware that you can only return one single HTML element from the render function. Meaning, there needs to be a single parent element such as a `div`. For example, you can't return two `h1` elements unless they are wrapped in a div element. You can return a single `h1` element as shown.

There are a few subtle differences in the HTML as it is used with React. These are as follows:

- The React DOM property naming convention is to use camel casing and has altered some of the standard properties slightly. For example, tabindex, becomes tabIndex and class becomes className.
- You use curly braces to embed JavaScript expressions in an attribute. For example, specifying the source image for an `<img>` element, you would have to do this: `src={myObject.someImageURL}`. You can use any object in your code this way.

To summarize, you can understand that a React application consists of provided Components that contain HTML to be rendered. This then is taken by the React API, and using the `document` object, React renders that into the actual browser DOM.

*Note: The React API also has calls for creating individual elements. For example, there is a method React.createElement() that takes arguments specifying the element tag, attributes and content. This is actually what the JSX gets turned into in the build process where the Babel compiler is run. This book will stick with using Components in JSX file syntax.*

## More about mixing JavaScript with markup

You briefly saw that in JSX syntax, you can have conditional code that uses control flow of JavaScript to decide what markup to return for rendering. This is really one of the great advantages of using React.

Imagine you had some backend service that you call to retrieve some data, such as from an HTTP/Rest web service. You can take that data and then process it in the JSX code and produce whatever UI you like.

The next example code will show the taking of some data that shows how code can make use of it. The data is a list of comments to be displayed. The code will first determine an appropriate piece of text to display and will then loop through the comments and display each using another lower level component. The looping code makes use of the JavaScript `map()` function of the array object to loop through and generate the HTML.

The CommentListItem is some custom component you would provide that would be reusable and know how to render an individual comment.

```
class Comments extends Component {
  render() {
    const coms = props.comments;

    return (
      {coms.length > 0 &&
        <h2>There are {coms.length} comments</h2>
      }
      <ul>
        {coms.map((comment, idx) =>
          <CommentListItem key={idx}
            title={comment.title} />
        )}
      </ul>

    );
  }
}
```

*Note:* You have the full power of pulling in JavaScript libraries like lodash that can be used to produce logic to output the markup you want rendered.

You can see what is possible in code that combines JavaScript code with HTML tags being returned. JavaScript can be inserted in the middle of HTML markup by placing it inside curly braces.

You can set up blocks of code that render based on a test, such as the one you see testing `coms.length`. Finally, you see the use of the `map()` function that lets you loop through and iterate to render a component. In this case, it is utilizing another component to create the li elements that would be placed inside the ul element to contain each list item.

*Note:* If you ever have logic that needs to decide between showing some UI or hiding it, it is better to not use CSS to hide the tags, but instead to just not to return anything at all. You simply return null from the code and that is it.

## Alternate syntax using a function

This book will always show components being built that are derived from Component. Theres is an alternate syntax that can be used that you should be aware of in case you ever see it.

Here is a component that is shown using the Component class syntax first and then is shown using the alternate syntax. One is using the arrow function syntax. All are consumed in the same way and have the props available to pass in. With the class usage, the props are of course part of the class instance.

```
// Code using a Component class
class Greeting1 extends React.Component {
  render() {
    return <h1>Greetings, {this.props.name}</h1>;
  }
}

// Code as a function
function Greeting2 (props) {
  return <h1>Greetings, {props.name}</h1>;
}

// With arrow function syntax
const Greeting3 = ({name}) => {
  <h1>Greetings, {name}</h1>;
}

// Usage
const g1 = <Greeting1 name="Joe" />;
const g2 = <Greeting2 name="Mary" />;
const g3 = <Greeting3 name="Paul" />;
```

There are limitations if you use the function syntax. For example, you cannot use Redux. You will also learn about what lifecycle events are, and find that these are only available if you use the Component class. You will only see code examples in this book using the `React.Component` class extension.

## 15.6 Custom Components and Props

You have seen a simple component named App. This was used in the initially created React application. It derived from the React Component class and had a `render()` method. Any user-defined component can act as a wrapper around the complexity of some underlying UI logic. Each component is really existing in order to render some HTML. You create components to split functionality out in a way that can be self-contained and reusable.

Components can be very complex and can have properties passed into them to control different aspects of their rendering. For example, you could pass in an array of data that the component then renders.

Components can maintain state, such as data about what is going on with the component. Passed in properties and internal state can be updated asynchronously and the UI gets rendered when the changes happen. React detects the changes, and updates will be batched to provide better performance.

# PART III: The Presentation Layer (React/HTML)

The capability of having properties on a component is what React calls "props". A component can have multiple props on it. Their usage looks just like that of attributes when you use a component in other markup. For example, you could pass in some text as a prop as shown in the following example. We can alter the index.js file to look as follows:

```
ReactDOM.render(
  <App name="Joe"/>,
  document.getElementById('root')
);
```

You can alter the App component as follows and then see it work. Here is the altered App.js file to use the passed in name. The text is taken and displayed along with the welcome text that specifies a person's name. The line in bold is the only modification needed.

```
// App.js
import React, { Component } from 'react';
import logo from './logo.svg';
import './App.css';

class App extends Component {
  render() {
    return (
      <div className="App">
        <div className="App-header">
          <img src={logo} className="App-logo" alt="logo" />
          <h2>Welcome to React {this.props.name}</h2>
        </div>
        <p className="App-intro">
          To get started, edit <code>src/App.js</code> and save to reload.
        </p>
      </div>
    );
  }
}

export default App;
```

Every prop shows up in the component as properties on the props object, and are available in the component class via `this.props`. The name property is accessed in the preceding example as this.props.name.

A property is only used as an input mechanism to a component. It is important to know that Props are immutable inside the component. For example, you could not have code that alters this.props.name. Props are meant to be passed in and consumed by the Component when it renders itself. The calling code is what sets the props and can change them if needed. If the calling code were to have a timer that later changed the name prop, then the UI would re-render.

You will learn that there are ways in the parent usage of a component where it can be notified of some change of data acquired by a child component. This is done by passing a callback function as a prop that the child component can use. This is totally up to you to define. There are predefined callbacks that can be used as standard HTML attribute notification mechanisms, such as `OnClick()`. These, however, are not surfaced back up to the parent, unless you take steps to provide the mechanism to feed it back up.

## 15.7 Components and State

As was mentioned, properties are passed into a component to be consumed via `this.props`. The other object available is `this.state`. State is internal to the component and can be changed internally. React knows about state changes and can refresh the UI to reflect any changes. Look at the following code and you will find the use the `this.state` object. Here is also where you will see the usage of a constructor for your component class.

```
class App extends Component {
  constructor(props) {
    super(props);
    this.state = { name: this.props.name };
  }

  render() {
    setTimeout(() => {this.setState(currState => ({
      name: "John"
    }))}, 5000);

    return (
      <div>
        <h2>Welcome to React {this.state.name}</h2>
      </div>
    );
  }
}
```

It is only in the constructor that you can make any direct assignment to `this.state`. In other places in your code, you call `this.setState()` to make changes. Also note that you call `super()`, which is the Component base class call for initializing its constructor. You can see in the code that the initial name property is set from the props that are passed in and then altered in a timer callback.

# PART III: The Presentation Layer (React/HTML)

Anytime you alter the state outside of the constructor, you must use the `this.setState()`. You only need to provide the properties you want to refresh in each case. It won't affect any of the other state properties of the component. React will do a UI refresh when it sees state changes have happened and may batch up several before it does a UI refresh.

With the usage of `this.setState()` you can access the current state before you make your change. In the previous example, we don't actually use that capability in the example.

The state should be moved up to the highest component which has it in common across child components. A parent can have a state property of its own and pass that into the child component as a prop. When the parent state changes, the child will see the update and React will re-render the component if necessary.

## Model, View and Controller

If you want to think in terms of how an MVC architecture works, you can recognize that React fits into this model by providing both the View and the Controller capability. The View is what renders DOM elements from your Component render method. The Controller is the rest of the logic that acts on behalf of the passed in props and managed state.

Your React component can do the fetching of the Model data. You have code fetch the data and pass it as props to a component, or in the component itself you fetch it and set the state. You can utilize libraries such as Flux/Redux to do the coordination of data fetching and the setting of state in the code of the component. Here is an image that might help you visualize how this all fits together in an MVC design:

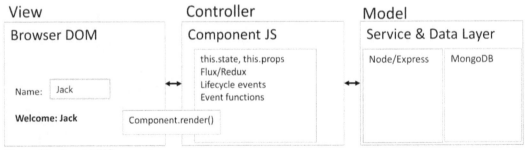

*Figure 92 - React MVC concepts*

**Note:** *You can debate exactly where to keep the controller and the view code. You could have it in one single component. It might be best, however, to separate the controller code from the view code. To do this you could have a data component that uses a data store and then the data component would pass props to a view only presentation component. These two types of components are called Container and Presentation Components (also called "smart" and "dumb"). More will be said about this later in the section on Redux usage.*

# 15.8 Event Handlers

The HTML tags used in rendering can have event handlers given to them as appropriate. This would be for events such as a click or key press. In React JSX files, this means you provide a JavaScript function in your React Component class and then use curly braces for specifying the event handler. Events are specified with camel case characters.

Here is an example that adds a handler for a click event on a button. Notice the arrow function is necessary to have the correct context for "this". Otherwise, the this.setState() would have an error thrown.

```
class App extends Component {
  constructor(props) {
    super(props);
    this.state = { count: 0 };
  }

  handleClick = (e) => {
    this.setState(currState => ({ count: currState.count + 1 }));
  }

  render() {
    return (
      <div>
        <h1>You have clicked the button {this.state.count} times</h1>
        <button onClick={this.handleClick}>
          Click it
        </button>
      </div>
    );
  }
}
```

The "e" function argument is actually an event object that has a lot of properties and methods on it to be used as needed. For example, you might want to capture a certain key that was pressed and then prevent it from propagating. There are cases where you don't want a click, or keypress to be processed any further. To prevent propagation of an event, you would call e.preventDefault().

## Passing up Event Handling

You can have the consumer of a custom component be notified of events in a sub-component by exposing them as properties. If you have multiple events, you need to name each event prop something different. In the code below, MyButton just calls the prop callback method in its `onClick` event. In this way, the App component can have the code that responds to the button click.

```
class App extends Component {
  constructor(props) {
    super(props);
    this.state = { count: 0 };
    // Alternate way to ensure correct this binding
    this.handleClick = this.handleClick.bind(this);
  }

  handleClick(e) {
    this.setState(currState => ({ count: currState.count + 1 }));
  }

  render() {
    return (
      <div>
        <h1>You have clicked the button {this.state.count} times</h1>
        <MyButton myClickHandler={this.handleClick}/>
      </div>
    );
  }
}

class MyButton extends Component {
  render() {
    return (
      <div>
        <button onClick={this.props.myClickHandler}>
          Click it
        </button>
      </div>
    );
  }
}
```

The code above also shows the alternate syntax for binding to have the correct calling on the callback, otherwise, you can use the arrow function syntax.

# Chapter 15: Fundamentals

# 15.9 Component Containment

There is a special props property that is always available on a component that is used for passing in child components or child HTML elements directly. The name of the property is "children". This means that you can enclose children in the usage of a custom component tag, and then get access to them. Here is the previous example, but this time, the text of the button is passed as part of the children.

```
class App extends Component {
  constructor(props) {
    super(props);
    this.state = { count: 0 };
  }

  handleClick = (e) => {
    this.setState(currState => ({
      count: currState.count + 1
    }));
  }

  render() {
    return (
      <div>
        <h1>You have clicked the button {this.state.count} times</h1>
        <MyButton myClickHandler={this.handleClick}>
          <h1>Click it</h1>
        </MyButton>
      </div>
    );
  }
}

class MyButton extends Component {
  render() {
    return (
      <div>
        <button onClick={this.props.myClickHandler}>
          {this.props.children}
        </button>
      </div>
    );
  }
}
```

The previous example shows an HTML tag being enclosed as part of the children. You can also have other JSX components or string literals be children as well.

*Note: When the JSX is taken and rendered, any whitespace at the beginning and end of a line is removed, as well as blank lines. New lines that occur in the middle of string literals are condensed down to a single space character.*

You can always pass in any name for a property that contains multiple elements. This may become necessary if you have multiple sub-components being used in elements and are displaying them. For example, in the previous example the button text could have been passed in as follows:

```
class MyButton extends Component {
  render() {
    return (
      <div>
        <button onClick={this.props.myClickHandler}>
          {this.props.myButtonText}
        </button>
      </div>
    );
  }
}
```

## 15.10 HTML Forms

React supports the creation of HTML forms via the standard form tag that exists in HTML. You can create a component that renders itself and provides the form tag and supports the event handling function for the submission. A simple example is as follow:

```
class App extends React.Component {
  constructor(props) {
    super(props);
    this.state = {value: ''};
  }

  handleChange = (event) => {
    this.setState({value: event.target.value});
  }

  handleSubmit = (event) => {
    alert('The test submitted is: ' + this.state.value);
    event.preventDefault();
  }
```

```
  render() {
    return (
      <form onSubmit={this.handleSubmit}>
        <h2>Text is {this.state.value}</h2>
        <label>
          Name:
          <input type="text" onChange={this.handleChange} />
        </label>
        <input type="submit" value="Submit" />
      </form>
    );
  }
}
```

Not only does the above code have the submission handling, but you also see code to handle the binding for changes of an input element so that the changes are rendered. You can also get events on changes for a checkbox, list, select control and others. The onChange property is used for those events.

## 15.11 Lifecycle of a Component

React sets up a series of stages that a component goes through as it gets rendered and eventually destroyed and taken out of the DOM. You can provide methods on your class that get called when the component enters each of these stages.

These methods are called "lifecycle hooks". For example, the `componentDidMount()` method is called after the component output has been rendered to the DOM. You can then know that your component is visible and take any action you need. You might want to create a timer that refreshes the state of the component every few seconds, if data changes in the backend services layer might have happened.

Another method available is `componentWillUnmount()`, which is called just before a component is taken out of the DOM.

## 15.12 Typechecking your Props

You can make use of the propTypes property of your component to set the expected property types of the usage of your component. This way, you can catch errors in its usage. This is only done when running in development mode.

You can test for individual props such as an object, array, func, string, number and many other types. You can also specify if a given prop is required or not. There is also a means of specifying default values for props that are not passed in by a parent. Here is an example that shows off these capabilities:

```
import PropTypes from 'prop-types';
class App extends Component {
  render() {
    return (
      <div className="App">
        <div className="App-header">
          <h2>Welcome to React {this.props.name1}</h2>
          <h2>Welcome to React {this.props.name2}</h2>
        </div>
      </div>
    );
  }
}

App.propTypes = {
  name1: PropTypes.string.isRequired,
  name2: PropTypes.string
};

App.defaultProps = {
  name2: 'Unknown person'
};
```

## 15.13 Getting a reference to a DOM element

There may be cases where you need to call the actual DOM manipulation functions on an element. To do that, you need to get a reference to be able to use it. React has a special syntax that you can employ to set the reference to the element. Here is what that looks like. You would do this if you can't store away a reference in your component class to an input element.

```
<input type="text" ref={(input) => { this.textInput = input; }} />
```

In some other code, you could make a call to use `this.textInput` to get or set its value. There is also a function provided in the React library named `findDOMNode` that can be used to find any element in code. I prefer to find ways around using either of these mechanisms as they really are there just for the odd case, as there usually is a way to find any component in context, and get its value in simpler ways.

# Chapter 16: Further Topics

That about covers the basics of using React. This chapter will cover further topics that you will want to know to achieve a full robust application.

## 16.1 Using React Router

It is assumed that a website will have many different pages to be viewed. In a SPA, this means many different views will be rendered as needed. To accomplish this you will need to render some type of navigation bar with links to click on. From those links the view would be changed.

React does not come with any built-in capability for handling page view routing. You could certainly create your own code to render some type of navigation bar across the top of your main site page and then do the work to change to different page views inside of that.

The solution I will give in this book is to utilize a popular open source npm package named React Router. There are actually two different packages you can find that are created by the same group of people. One is for websites and the other is for native app development. We will use the website version, which is found here - https://www.npmjs.com/package/react-router-dom. In case you are interested, the native app version is named react-router-native. React Router gives you the ability to create a dynamic navigation experience on your site. To get started, you can install the package into your project as follows:

```
npm install --save react-router-dom
```

*Note: React Router can be utilized on the server-side to return pages that are rendered with data, or React Router can be used a pure client-side SPA site. This book will focus on the client-side SPA usage and show how views are populated from client-side fetches of data from the server HTTP/Rest API created in part two of this book.*

You can put at the top of your App.js file an import statement to pull in the needed components that are provided by the react-router-dom package. These would be `HashRouter`, `Route` and `Link`. Here is the code that could be placed into App.js for a simple usage that has no CSS styling for it:

```
// App.js
import React, { Component } from 'react';
import './App.css';
import { HashRouter, Route, Link } from 'react-router-dom'
```

## PART III: The Presentation Layer (React/HTML)

```
const Home = () => (
  <div>
    <h2>Home</h2>
  </div>
)

const About = () => (
  <div>
    <h2>About</h2>
  </div>
)

class App extends React.Component {
  render() {
    return (
      <HashRouter>
        <div>
          <Link to="/" replace>Home</Link>
          <Link to="/about" replace>About</Link>
          <hr />
          <Route path="/about" component={About} />
          <Route exact path="/" component={Home} />
          <Route component={NotFound} />
        </div>
      </HashRouter>
    );
  }
}
export default App;
```

The `HashRouter` is the main component wrapper that creates the container for the UI to be rendered as navigation happens. Different components are rendered as assigned by the "component" attribute of the Route. For a site that you want browser history kept, use BrowserRouter instead of HashRouter. In the above case, we are discussing a SPA, which would be implemented with the hash routing.

The attribute "exact" is used to match only a single specific path. If you have multiple components being rendered at a time, this is needed and will fix that problem.

The Link component is used as an anchor tag might also be used, but adds the prop "to" for specifying the route to transition to. The Link component works hand-in-hand with the Route component which specifies what component to render for a given route. The "replace" attribute of the Link specifies that clicking the link will replace the current entry in the history stack instead of adding a new one. This is what we want for a SPA client-side site.

## Chapter 16: Further Topics

The last Route listed is one that lets all other paths be rendered with an error. If you end up there, you would provide a NotFound component to state that this is a "404", because it was not expected. For example, this NotFound UI would render if you went to something like localhost:3000/#/blahblahblah.

Here is what the UI ends up looking like for this previous example code:

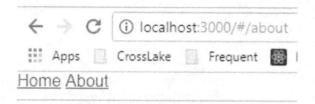

*Figure 93 - React Router usage*

## 16.2 Using Bootstrap with React

Styling a website to create a modern responsive site that looks good in a full-screen browser as well as on a small mobile phone is a challenge. There is much to consider when constructing your HTML in this case. This book is not a way for you to learn CSS or responsive Web Design techniques. What this book will do instead is take a simple and effective approach by using Bootstrap.

Bootstrap was created by Twitter and is a set of UI HTML templates and CSS styles that you can use to accomplish responsive website designs.

To make this even easier in a React application, there is an npm package that exports React Components to be easily used. You just need to run the following install to get everything available in the project:

```
npm install --save react-bootstrap bootstrap@3
```

The only other change to make is to pull in the CSS files needed so that they are exposed for usage in the application. Here is the altered index.js file that is found in the src directory. The added lines are in bold.

# PART III: The Presentation Layer (React/HTML)

```
// index.js
import React from 'react';
import ReactDOM from 'react-dom';
import App from './App';
import 'bootstrap/dist/css/bootstrap.css';
import 'bootstrap/dist/css/bootstrap-theme.css';
import './index.css';

ReactDOM.render(
  <App/>,
  document.getElementById('root')
);
```

Here is an altered App.js file that now shows the usage of a Bootstrap styled button:

```
// App.js
import React, { Component } from 'react';
import { Button } from 'react-bootstrap';

class App extends Component {
  render() {
    return (
      <div>
        <h1>Welcome to React with Bootstrap</h1>
        <p><Button
            bsStyle="success"
            bsSize="large"
            href="http://react-bootstrap.github.io/components.html"
            target="_blank">
            View React Bootstrap Docs
        </Button></p>
      </div>
    );
  }
}
export default App;
```

Here is what the rendering of the example above would look like:

*Figure 94 - Bootstrap usage with React*

## Chapter 16: Further Topics

You can visit this React-Bootstrap site (https://react-bootstrap.github.io/components/alerts/) to see all of the available components you can use in your React application. The NewsWatcher application will make use of several of them.

# 16.3 Making HTTP/Rest Requests

Every application that you create will most likely need to make HTTP/Rest requests to some back-end service to retrieve data. React does not have any such capability built in to do that. This was done on purpose, as there are several libraries that are already good at this. One option is to use the, built in standard of `fetch()` as found in the JavaScript standard. Other npm modules such as superagent can be installed and used. This book will utilize `fetch()`.

Typically, you will need to fetch data in the `componentDidMount()` method. That code will update some internal state properties and in turn would cause React to do any DOM updates necessary. Here is an example that outputs the fetched data to the console log:

```
import superagent from 'superagent';
...Code left out...
componentDidMount() {
  fetch(`/api/blah`, {
    method: 'GET',
    headers: new Headers({
      'x-auth': this.props.session.token
    })
  })
  .then(r=>r.json().then(json=>({ok:r.ok, status:r.status, json})))
  .then(response => {
    if (!response.ok || response.status !== 200) {
      throw new Error(response.json.message);
    }
    console.log(response.json);
  })
  .catch(error => {
    console.log(`GET failed: ${ error.message }`);
  });
}
```

This type of code will be used in when NewsWatcher is put together. The basic usage is that you specify the URI or relative path. As part of the second parameter object, you give the HTTP verb usage, such as put, post, get or delete. You can also set any headers you need. Fetch can be used as promise based code, and that is what is shown here. You can verify the HTTP code returned and handle errors and get at the body of the returned data.

PART III: The Presentation Layer (React/HTML)

# 16.4 State management with Redux

You have learned how to use props and state. State and sometimes props that are shared should not be kept as part of a component class, but instead should be kept in a central storage location for all components to get at. If you find that the props or state of a component really needs to be shared with other components, they should be stored centrally.

Redux is a library that helps you share state across all your React components. Its purpose is to provide a central repository for all components to write to and receive updates from. Think of it as fulfilling the same function that state does in a component, but across all components in a shared way. With Redux, you can also accomplish data transformations and routing in the process.

The state that you store can be some data you collected from user interaction, or data that comes from a network request and much more. Just about any type of simple JavaScript data as part of an object can be kept there.

Imagine retrieving data that needs to be shared across the hierarchy of components. For example, the UI in one part of the component hierarchy might have state that needs to be shared with any number of other components strewn all over in the hierarchy. It would get confusing if you tried to send that state up, across and down to another component. That would quickly become incomprehensible.

Redux solves these problems by being the central state repository for your complete hierarchy across all components. To use Redux, you first need to create the Redux store. You do this in your code where it is initialized. When you create the store, you pass into it the reducers. A reducer is what gets a chance to determine the state changes you want to make when data flows in the form of an action. This reducer could be a single function, or a combination of reducers. This will become clearer as you look at some code examples.

The npm package that you install is called "redux". The Redux object that you can access has only four methods on it, so it is fairly easy to understand. There are then just a few nuances to understand about each of the four calls. Here is the creation of a store and also some code showing the four Redux functions:

```
import { createStore } from 'redux'
const store = store = createStore(reducer);
store.dispatch(action)
store.subscribe(listener)
store.getState()
store.replaceReducer(nextReducer)
```

## Chapter 16: Further Topics

The createStore() call needs one or more reducers given to it. It can also be passed some middleware. There are other npm packages you can download to act as middleware, or you can write your own. This just means that the flow can be intercepted and altered in some way. This is similar to what middleware does in Express with Node.js. I have not needed to use this capability.

The dispatch() call is what you use to cause a change in state to happen. The getState() function is a way to fetch the state at that moment in time and the subscribe() is for receiving asynchronous notifications as state is changed.

In the NewsWatcher codebase, I make the initial call to createStore() with some provided reducer functions and then call dispatch() a lot. Tere is also usage of another module on top of Redux that I will soon describe.

## The flow of a state change

What you would typically do in developing your React code is to code up a component and use this.state for keeping data that changes local to that component. You might recall that you do that in a component with the this.setState() call. Then as you progress and find that the state is something that needs to be shared, you move it to the Redux state storage. Don't put everything in Redux as that can become overwhelming.

To update the Redux state, you make a call to store.dispatch(action). For example, if a component wanted to update some state and make that available back to itself and to other components, it would make that call. The component calling dispatch() does not need to know about who will listen for that state change. This is a single direction of flow with the state change. There are no bi-directional messages flowing through Redux in any kind of conversation.

The simple usage is to preconfigure Redux with what are called reducers. These act as a way to filter data as it flows through. Dispatch calls happen and flow through what is called an "action" in the reducers which then causes the particular affected state to change and be available.

Code receives the changes in the state via a store.subscribe() call. You can also get a snapshot of the state at a given instant with the call store.getState(). Here is a visual representation of flow for a state change as it passes via the Redux store:

# PART III: The Presentation Layer (React/HTML)

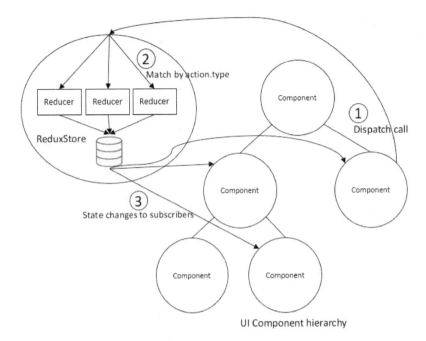

*Figure 95 - React state management*

Here is the simplest code that can be used to demonstrate the basic flow. Note that the action object must at least have a property named "type". This is used to find the correct router to use for processing. The other properties of the action object are up to you to pass and deal with. The following code creates the Redux store with the passed in reducer. The function usage has an initial object passed in that sets the storage of the state property.

```
const initialState = {
   count: 0,
   currentMsg: "Hello Redux"
}

const startState = (state = initialState, action) => {
   switch (action.type) {
     case 'INCREMENT':
       return {
         ...state,
         count: state.count + action.incAmount;
       }

     default:
       return state;
   }
}
```

## Chapter 16: Further Topics

```
const { createStore } = Redux;
const store = createStore(startState);

const Counter = ({
  value,
  onIncrement
}) => (
  <div>
    <h1>How many bugs have you fixed today?</h1>
    <h1>{value}</h1>
    <button onClick={onIncrement}>+</button>
  </div>
);

const render = () => {
  ReactDOM.render(
    <Counter
      value={store.getState().count}
      onIncrement={ () => store.dispatch({type: 'INCREMENT',
incAmount: 1}) }
    />,
    document.getElementById('root')
  )
};

store.subscribe(render);
render();
```

The state for a given reducer can take any form you define for it. You can see in the previous code that it is defined with default values to be initialized with. Redux makes an initial call to the reducer on its own and the defaults are set.

To alter the state, a dispatch call happens and passes in the action object with the type and values to consume. In the Redux code, the state is immutable, so you cannot change it directly, you have to make a clone of it (i.e. see JavaScipt `Object.assign()` usage, such as `Object.assign({}, state, { users: action.users });`), or use something like the immutability-helper. This will be used in the NewsWatcher sample application. You also need to return the complete state object back, as it will all be replaced. The code above simply creates the new object from properties of the existing state using the object spread syntax of Javascript. Then the count is updated.

For example, if it was an object, you replace the whole thing, not just one property. With the what you saw with React component state, you have one object that had independently updatable properties on it. Redux does not work the same way. You can, however, have different reducers that are independent. Updating one, does not affect the other.

297

PART III: The Presentation Layer (React/HTML)

Most applications will want to split out what is stored in the Redux storage into separate concerns. You simply do this by calling `combineReducers()` to piece them all together. Each one will receive all of the actions routed to them, but it is of course up to the reducers to match what `action.type` values they care about. You call createStore with this combined rootReducer instead of a single reducer function.

```
import { combineReducers } from 'redux'

const rootReducer = combineReducers({
  app,
  news,
  sharednews,
  profile
})
```

## Usage of Redux with react-redux

The Redux library is a reusable JavaScript library that can be used in JavaScript code. It is certainly usable as it is with is four methods. There are, however, certain code design patterns of React that are used over and over. One of these patterns involves state manipulations. In order to combine that pattern with the use of Redux, in an easy to consume way, there was a library created named react-redux.

To get started, you need to install the npm packages redux and react-redux. You then need to add some code to create the single Redux store. There is a special React component that is used to wrap your application component so as to make Redux available in all of the component hierarchy. Here is what the index.js file ends up looking like. The key changes have been highlighted in bold font.

```
import React from 'react';
import ReactDOM from 'react-dom';
import { createStore } from 'redux'
import { Provider } from 'react-redux'
import reducer from './reducers'
import App from './App';
import 'bootstrap/dist/css/bootstrap.css';
import 'bootstrap/dist/css/bootstrap-theme.css';
import './index.css';

const store = createStore(reducer);
ReactDOM.render(
  <Provider store={store}>
    <App />
  </Provider>, document.getElementById('root')
);
```

298

## Chapter 16: Further Topics

The App component will be a child of the Provider component. The code from react-redux library will be able to use it and pass in additional props. It will pass in a prop that is a function named "dispatch". This way, you don't need to pass the Redux store to the App component. You won't actually be coding up calls to `store.dispatch()`, just `dispatch()`.

The next thing you need to understand is how to take a component you write and do the work to get the state from the changes that happen in the Redux store and transfer that state to the local props on the component.

The state coming from the Redux store can be used in the props on a component. Props are then changed via the dispatch to cause the behavior to change. You do not take Redux state and transfer it to the `this.state` of a component. That would be a waste, as they both serve the same purpose and would be duplicated and you would not have a single source of truth.

Here is what you need to do in a component to have it pull in state changes from a Redux store:

```
import { connect } from 'react-redux'

...lots of component code left out...

MyComponent.propTypes = {
  dispatch: PropTypes.func.isRequired
};
const mapStateToProps = state => {
  return {
    session: state.app.session,
    user: state.profile.user,
    isLoading: state.profile.isLoading
  }
}
export default connect(mapStateToProps)(MyComponent)
```

This code may seem a little strange when you first look at it. What you are doing here is instead of a simple export of `MyComponent`, you call the react-redux provided `connect()` function and that has the ability to be able to take your component and wrap it for its own purposes. Redux keeps track of when it needs to send state updates and re-render it and provide the dispatch prop. You can see that you can get whatever you like from the Redux store. This time it is pulling from the app and profile settings.

You pass to the `connect()` call a callback function that lets you get access to the Redux state. This function is called every time anything in the Redux state changes. Using the state

## PART III: The Presentation Layer (React/HTML)

passed in, you can take whatever your particular component cares about. This contains all of the different stored state from all of the reducers that are in use.

There is more that is possible with react-redux. For example, you can pass other parameters to `connect()`, such as a function that allows you to hook up dispatch calls from your component, such as UI click handlers that make dispatch calls. Then somewhere in your code, you can call the prepared function that is available on the props.

```
import { connect } from 'react-redux'

...lots of component code left out...

sendAction = () => {
    this.props.sendAction()
}

render() {
    <div>
      <Button onClick={sendAction}/>
    </div>
}

MyComponent.propTypes = {
  dispatch: PropTypes.func.isRequired
};

const mapStateToProps = state => {
  return {
    session: state.app.session,
    user: state.profile.user,
    isLoading: state.profile.isLoading
  }
}


function mapDispatchToProps(dispatch) {
    return({
        sendAction: () => {dispatch(
            { type: 'SOME_ACTION', someData: 1 })}
    })
}

export default connect(mapStateToProps,
             mapDispatchToProps)(MyComponent)
```

## Chapter 16: Further Topics

You don't have to set things up this way, as `dispatch()` is available anyway as a prop of the component already. Here is an example that adds in the `mapDispatchToProps` function:

If you don't want to set up a `mapDispatchToProps` then you just call dispatch from `this.props` as follows.

```
sendAction = () => {
    this.props.dispatch({ type: 'SOME_ACTION', someData: 1 })
}
```

*Note: The usage of additional libraries like Redux is completely at your discretion. Only use a library if at any point you see that you can simplify your code. Sometimes a simple code base with no design frills works just fine. You must decide when and how to refactor and introduce new libraries into your code and what the cost versus benefit will be. In the case of Redux, I believe there is a clear benefit. You will see it used throughout the NewsWatcher code.*

# Chapter 17: NewsWatcher App Development with React

With the previous chapter's content, you are now able to understand the construction of the NewsWatcher presentation layer code. If you followed along in the previous chapter in the section on installation steps, you are all set to begin writing code.

In this chapter, you will learn about the files as they exist in the project posted on Github and how each one is pieced together. If you clone the project from GitHub, the project folders will look as follows:

*Figure 96 - VS Code file tree*

**Note**: *Don't forget that you can access all of the code for the sample NewsWatcher project at https://github.com/eljamaki01/NewsWatcher2RWeb.*

## Chapter 17: NewsWatcher App Development with React

You may have noticed that you had a placeholder index.html file that your Node.js service was serving up. The existing Node.js project from part two of this book is capable of serving up your UI after you make just a few minor tweaks.

# 17.1 Where it All Starts (src/index.js)

Back in the sections on the Node.js web service development, you saw a few lines of code in the server.js file that served up a `get` request for the main HTML page. This allowed for the downloading and display of the NewsWatcher site.

The express routing for that file was separated from those providing access to the backend API. Here are the lines from server.js for the route handling that serves up the index.html file and its associated static resources:

```
app.get('/', function (req, res) {
  res.sendFile(path.join(__dirname, 'build', 'index.html'));
});
app.use(express.static(path.join(__dirname, 'build')));
```

The `app.get()` call gives a specific route for a `get` request to your root site and states that you will always serve up your index.html file for that. This is really all you need to set up the use of React to get the react application sent to be run on a client-side browser.

The app.use() call sets up middleware for a router path for all of the static files. You simply tell it that you have this directory named "build" for where they all are located.

You will note that this final index.hml file is not a file you created, but it is generated for you in the build process from the index.js file that you provided and the template index.html file found in the public folder. Look in the build folder to find the generated index.html file that gets served up.

Here is the index.js file. The few additions you would find here are just the inclusion of css files that can be used across anything that is rendered. In particular, you see the ones for styling bootstrap elements. You also see the code to set up Redux to be used across all the application. The reducers are brought in and the Provider component is something from react-redux that provides Redux to the overall application.

```
import React from 'react';
import ReactDOM from 'react-dom';
import { createStore } from 'redux'
import { Provider } from 'react-redux'
import reducer from './reducers'
```

303

## PART III: The Presentation Layer (React/HTML)

```
import App from './App';
import 'bootstrap/dist/css/bootstrap.css';
import 'bootstrap/dist/css/bootstrap-theme.css';
import './index.css';

const store = createStore(reducer);

ReactDOM.render(
  <Provider store={store}>
    <App />
  </Provider>, document.getElementById('root')
);
```

The call to `ReactDOM.render()` is the starting point call to react that renders to the root div element. All of your UI emanates from here, and comes from the App component.

# 17.2 The hub of everything (src/App.js)

The App.js file is where we establish the UI that renders the views. This means that it provides things like the capability to present a navigation bar and allows for the navigating between the different page views, such as logging in, viewing news, and setting profile settings. This file sets up the App class that is derived from the Component class.

**Imports, constructor and the componentDidMount lifecycle**
The imports are set up to bring in external libraries and the other components that are needed for views that are used in the navigation. Then there is the code for the App component class.

Redux is used for the state settings that are needed. You will find the use of the `dispatch()` function to store state data in Redux in a central way that is global to the application. Look at the `mapStateToProps()` function at the bottom and you will see what properties the App class retrieves from Redux. They are:

- The state object to tell us if we are signed in.
- The session token that was retrieved from the server-side log in.
- The status message that is displayed at the top of the UI.

The `componentDidMount()` lifecycle event method is where you do processing once the UI has been rendered and need to make any alterations. The code checks the local browser storage to see if a token has been saved away to be retrieved and if so gets that and sets a message that the user is logged in. The user would need to have selected the option at login time to have one saved. The dispatch that happens places that token into Redux storage to global access from all other pages that need it, such as for profile retrieval.

## Chapter 17: NewsWatcher App Development with React

The logged in flag is set to true in that same dispatch processing, which alters the menu rendering, as you will soon see. The value of `currentMsg` is set through the dispatch and would show up in the UI.

The UI is rendered once from the server, with the initial HTML, with its CSS and JavaScript and then it is completely self-sufficient in the client browser from then on. There are no pages being rendered from the server side after that. You could create a combination of server and client-side page rendering in a true isomorphic application as will be discussed later. Here is the code that has been discussed thus far.

```
//App.js
import React, { Component } from 'react';
import './App.css';
import { HashRouter, Switch, Route } from 'react-router-dom'
import { Navbar, Nav, NavItem } from 'react-bootstrap';
import { IndexLinkContainer } from 'react-router-bootstrap';
import { connect } from 'react-redux'
import LoginView from './views/loginview';
import NewsView from './views/newsview';
import HomeNewsView from './views/homenewsview';
import SharedNewsView from './views/sharednewsview';
import ProfileView from './views/profileview';
import NotFound from './views/notfound';

class App extends Component {

  componentDidMount() {
    // Check for token in HTML5 client side local storage
    const storedToken = window.localStorage.getItem("userToken");
    if (storedToken) {
      const tokenObject = JSON.parse(storedToken);
      this.props.dispatch({ type: 'RECEIVE_TOKEN_SUCCESS', msg: `Signed in as ${tokenObject.displayName}`, session: tokenObject });
    } else {
    }
  }

...code left out...
}

const mapStateToProps = state => {
  return {
    loggedIn: state.app.loggedIn,
    session: state.app.session,
    currentMsg: state.app.currentMsg
  }
}

export default connect(mapStateToProps)(App)
```

305

# PART III: The Presentation Layer (React/HTML)

This application is a SPA application, so hash based browser navigation is used. This does not allow for history to be kept and if you try and use the browser back button, the application would not have any history to go back and forth from.

**Navigation bar**
The main point of this App component is to set up the navigation bar that will exist for users to get access to the menus and see a status message. The DOM rendering makes use of some handy bootstrap styling. You can investigate the particulars of how it all works on your own through the usage of the react-bootstrap npm package. Here is an image of what would be rendered on a smartphone device. It shows the UI state when the menu is opened.

*Figure 97 - Sample image from a smartphone*

The react-boostrap usage has the ability to do what is called "responsive" web rendering and it creates a button that will be used to provide a drop-down menu when it is being run on a mobile smartphone. If you run it in a desktop browser, it does not appear this way, but the navigation bar is spread out. Bootstrap classes adapt to the size of the display.

***Note:*** *The great thing is that you don't need to use the raw bootstrap stylings with low-level HTML elements. React-boostrap wraps all of that and presents it for your usage in react Components that are consumed.*

As part of the header of the navbar, there is a span element used to display messages you want the user to see. For example, if their login fails, you want the user to know that. The state is set through Redux with the appropriate message string.

Each `IndexLinkContainer` entry represents the menu selections for navigating around the application. There is a clever way to show or hide each entry depending on the state of the app. The `loggedIn` property is used to determine the display of navigation bar entries. When a user is logged in, you want all of the menu selections visible. Until then, they are hidden.

## Chapter 17: NewsWatcher App Development with React

The actual navigation click-handling is done through the use of the react-router-dom npm package. At the very top of what is rendered, is the usage of the `HashRouter` component. This is what wraps everything and gives us the ability to render in the DOM whatever we set up as a page view route. Then we can either let the navigation bar selections control the route to select, or in some cases, manage the navigation programmatically.

You can see the react-router-dom Switch component that is used in conjunction with the Route component to handle what page views get rendered. The interesting thing is that for each Route, you specify the path that it is for and the component to render. In one case, the code has to actually use the render Prop in order to pass the state into the usage of the page view component. Here is the `render()` function for the App component that controls the navigation:

```
render() {
  return (
    <HashRouter>
      <div>
        <Navbar fluid default collapseOnSelect>
          <Navbar.Header>
            <Navbar.Brand>
              NewsWatcher {this.props.currentMsg &&
<span><small>({this.props.currentMsg})</small></span>}
            </Navbar.Brand>
            <Navbar.Toggle />
          </Navbar.Header>
          <Navbar.Collapse>
            <Nav>
              {<IndexLinkContainer to="/" replace><NavItem >Home Page News</NavItem></IndexLinkContainer>}
              {this.props.loggedIn &&
                <IndexLinkContainer to="/news" replace>
                  <NavItem >My News</NavItem>
                </IndexLinkContainer>}
              {this.props.loggedIn &&
                <IndexLinkContainer to="/sharednews" replace>
                  <NavItem >Shared News</NavItem>
                </IndexLinkContainer>}
              {this.props.loggedIn &&
                <IndexLinkContainer to="/profile" replace>
                  <NavItem >Profile</NavItem>
                </IndexLinkContainer>}
              {this.props.loggedIn &&
                <NavItem onClick={this.handleLogout}>Logout</NavItem>}
              {!this.props.loggedIn &&
                <IndexLinkContainer to="/login" replace>
                  <NavItem >Login</NavItem>
                </IndexLinkContainer>}
            </Nav>
          </Navbar.Collapse>
        </Navbar>
```

## PART III: The Presentation Layer (React/HTML)

```
        <hr />
        <Switch>
          <Route exact path="/" component={HomeNewsView} />
          <Route path="/login" component={LoginView} />
          <Route path="/news" component={NewsView} />
          <Route path="/sharednews" component={SharedNewsView} />
          <Route path="/profile" render={props => <ProfileView
appLogoutCB={this.handleLogout} {...props} />} />
          <Route component={NotFound} />
        </Switch>
      </div>
    </HashRouter>
  );
}
```

*Note: Look at the top of the top of the file App.js file to see the import statements. These will help you keep straight which components come from which npm packages. For example, those from react-bootstrap and those from react-router-dom.*

### Logging out

The navigation bar has a selection to log the user out. This simply places a call to the backend service and it can do whatever it would like code wise, but the client side really just needs to set the state to reflect that and redirect the view to the login. This is done through a browser capability with the use of setting `window.location.hash`.

```
handleLogout = (event) => {
  const { dispatch } = this.props
  event && event.preventDefault();
  fetch(`/api/sessions/${this.props.session.userId}`, {
    method: 'DELETE',
    headers: new Headers({
      'x-auth': this.props.session.token
    }),
    cache: 'default' // no-store or no-cache?
  })
  .then(r => r.json().then(json => ({ ok: r.ok, status: r.status, json })))
  .then(response => {
    if (!response.ok || response.status !== 200) {
      throw new Error(response.json.message);
    }
    dispatch({ type: 'DELETE_TOKEN_SUCCESS', msg: "Signed out" });
    window.localStorage.removeItem("userToken");
    window.location.hash = "";
  })
  .catch(error => {
    dispatch({type:'MSG_DISPLAY',msg:`Sign out failed: ${error.message}`});
  });
}
```

Chapter 17: NewsWatcher App Development with React

# 17.3 Redux Reducers (src/reducers/index.js)

You saw in the src/index.js file how the reducers were brought into the application when it started up. The line was as follows that set up the Redux Store:

```
import reducer from './reducers'
const store = createStore(reducer);
```

In case you forgot, if you have a require or import statement of a directory, it will by default look for a file named index.js and use that.

Index.js then combines all the separate reducers into one and exports that:

```
// src/reducers/index.js
import { combineReducers } from 'redux'
import app from './app'
import homenews from './homenews'
import news from './news'
import sharednews from './sharednews'
import profile from './profile'

const rootReducer = combineReducers({
  app,
  homenews,
  news,
  sharednews,
  profile
})

export default rootReducer
```

Let's now look at a few of the reducers bring combined.

**The App Reducer**
The App reducer has three actions that it supports – `MSG_DISPLAY`, `RECEIVE_TOKEN_SUCCESS` and `DELETE_TOKEN_SUCCESS`. We have seen these used in the src/app.js file where the redux state is used for each of these. In the case of the `MSG_DISPLAY`, it simply sets a new string for the text of the `currentMsg` property.

Notice the "`...state`" code that is a JavaScript way of taking an object and pulling in all of its properties. This way, you don't have to list them. This way you get whatever was set for session and loggedIn and then `currentMsg` is overridden.

```
// src/reducers/app.js
const initialState = {
  loggedIn: false,
  session: null,
  currentMsg: ""
}

const appLevel = (state = initialState, action) => {
  switch (action.type) {
    case 'MSG_DISPLAY':
      return {
        ...state,
        currentMsg: action.msg
      }
    case 'RECEIVE_TOKEN_SUCCESS':
      return {
        ...state,
        loggedIn: true,
        session: action.session,
        currentMsg: action.msg
      }
    case 'DELETE_TOKEN_SUCCESS':
      return {
        ...state,
        loggedIn: false,
        session: null,
        currentMsg: action.msg
      }
    default:
      return state
  }
}

export default appLevel
```

I will show one more reducer. This one will illustrate the point that the state in the Redux store is immutable, so you have to either replace the complete object, or use a special module to help you manage that.

There is an npm module named immutability-helper that allows you to make a change to an immutable object. Part of this state setting is a news property that is an array of news stories with their comments.

A new comment is added with the ADD_COMMENT_SUCCESS action. It does that by pushing it to the end of the array for a news story. In other words, there is an array of news stories and each of those have an array of comments. If we did not use the immutability helper, a completely new copy of the array would have to be made and replaced each time.

## Chapter 17: NewsWatcher App Development with React

```javascript
// src/reducers/sharednews.js
import update from 'immutability-helper';

const initialState = {
  isLoading: true,
  news: null
}

const news = (state = initialState, action) => {
  switch (action.type) {
    case 'REQUEST_SHAREDNEWS':
    return {
      isLoading: true,
      news: []
    }
    case 'RECEIVE_SHAREDNEWS_SUCCESS':
    return {
      // ...state,
      isLoading: false,
      news: action.news,
    }
    case 'ADD_COMMENT_SUCCESS':
    return {
      ...state,
      news: update(state.news, {
        [action.storyIdx]: {
          comments: {
            $push: [{ displayName: action.displayName, comment: action.comment }]
          }
        }
      })
    }
    default:
      return state
  }
}

export default news
```

The immutability helper syntax is a bit tricky here. What is being altered is an array property found in an object in the new array. The first thing to do is select the index of the story in the new array, and then to push a comment to the comments array found in that object entry.

A less efficient way would be to clone the array and then alter it. Some array functions like `slice()` actually return a new array and don't mutate the old one. You can thus get the new array and then alter it as you like.

## PART III: The Presentation Layer (React/HTML)

You can use the JavaScipt `Object.assign()` that allows you to create a new object from the existing state object and then override the comments property of that. Here is how the code would be for an array using the `slice()` function:

```
case 'ADD_COMMENT_SUCCESS':
   var newNews = state.news.slice(0);
   newNews[action.storyIdx].comments.push({ displayName: action.displayName, comment: action.comment });
   return {
     ...state,
     news: newNews
   }
```

The rest of the sections in this chapter go over the code that exists in each of the views that are rendered – login, homeNews, news, shared news and profile. In the previous chapters on React fundamentals, you learned about Redux and refactoring your component to split them into controller and view components. I have chosen not to complicate the code by splitting out each view into its controller/container component. I keep all the code in one single component.

## 17.4 The Login Page (src/views/loginview.js)

To log a user in, their email and password are entered and verified by the backend service. The login page contains a checkbox for the user to specify that they want their local device to store their login token for them. If the user is not registered yet, they can click to bring up a popup modal dialog form to register a new account. As part of the registration, an email needs to be provided. Emails are unique in the system, so no duplicates are allowed in the data layer. Here is what the UI looks like for logging in.

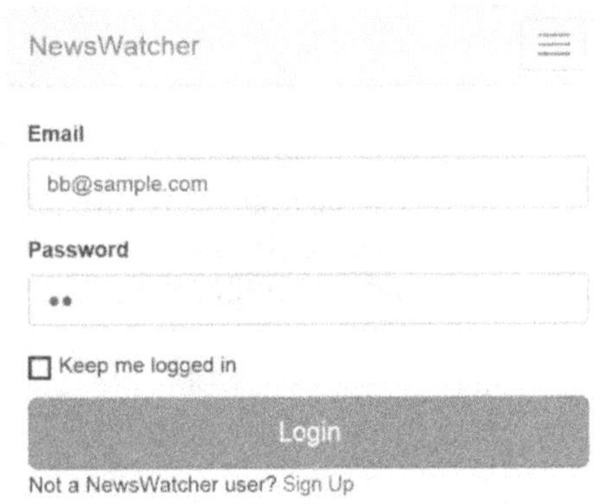

*Figure 98 - NewsWatcher login form*

## Chapter 17: NewsWatcher App Development with React

The HTML code for the form is set up to have a submit handler. The `handleLogin()` function of the component is the code that executes at that time. To get the data from the form, various Bootstrap components are used. These bind their data to the state of the component through `onChange` handlers that get set as each character is typed. Here is the render code for the login page as found in the loginview.js file:

```
render() {
  // If already logged in, don't go here and get routed to the news view
  if (this.props.session) {
    return null;
  }
  return (
    <div>
      <form onSubmit={this.handleLogin}>
        <FieldGroup
          id="formControlsEmail2"
          type="email"
          glyph="user"
          label="Email Address"
          placeholder="Enter email"
          onChange={this.handleEmailChange}
        />
        <FieldGroup
          id="formControlsPassword2"
          glyph="eye-open"
          label="Password"
          type="password"
          onChange={this.handlePasswordChange}
        />
        <Checkbox checked={this.state.remeberMe}
          onChange={this.handleCheckboxChange}>
          Keep me logged in
        </Checkbox>
        <Button bsStyle="success" bsSize="lg" block type="submit">
          Login
        </Button>
      </form>
      <p>Not a NewsWatcher user?
        <a style={{ cursor: 'pointer' }}
          onClick={this.handleOpenRegModal}>Sign Up</a>
      </p>
      {this._renderRegisterModal()}
    </div>
  );
}
```

You can see that at the top of the `render()` function we first see if we are logged in. When we are logged in, the session property from Redux is set.

PART III: The Presentation Layer (React/HTML)

The other very interesting thing that is going on is that you see the call to the function `renderRegisterModal()`. I could have just placed all the HTML right there in the render function. I did not do that as that function is large enough already. Breaking it out makes the code easier to read. This takes that out and makes it self-contained.

The other thing to know is that this UI for registering a user of NewsWatcher is a modal dialog, and uses the `<Modal>` component that is provided by Bootstrap for React. It is always rendered, but it is being shown or hidden by a state property and you can see that it starts out as being hidden.

The modal registration dialog UI looks as follows:

*Figure 99 - NewsWatcher Registration Modal*

Here is the code for the registration modal UI:

```
_renderRegisterModal = () => {
  return (<Modal show={this.state.showModal}
onHide={this.handleCloseRegModal}>
     <Modal.Header closeButton>
       <Modal.Title>Register</Modal.Title>
     </Modal.Header>
     <Modal.Body>
       <form onSubmit={this.handleRegister}>
```

```
        <FieldGroup
          id="formControlsName"
          type="text"
          glyph="user"
          label="Display Name"
          placeholder="Enter display name"
          onChange={this.handleNameChange}
        />
        <FieldGroup
          id="formControlsEmail"
          type="email"
          glyph="user"
          label="Email Address"
          placeholder="Enter email"
          onChange={this.handleEmailChange}
        />
        <FieldGroup
          id="formControlsPassword"
          glyph="eye-open"
          label="Password"
          type="password"
          onChange={this.handlePasswordChange}
        />
        <Button bsStyle="success" bsSize="lg" block type="submit">
          <Glyphicon glyph="off" /> Register
            </Button>
      </form>
    </Modal.Body>
    <Modal.Footer>
      <Button bsStyle="danger" bsSize="default"
onClick={this.handleCloseRegModal}><Glyphicon glyph="remove" />
Cancel</Button>
    </Modal.Footer>
  </Modal>)
}
```

This UI uses a form and has a function that handles the submit. Let's now look at the code that supports the LoginView component.

### Component supporting code of LoginView

There is the usual constructor that sets up what is needed for the state. In this case, we have properties that will hold the bound data from the form, such as email and password. The `showModal` property is used to show and hide the modal registration form.

As explained, at the top of the render function is a test to see if there is a session token and a null is returned, which React sees and ignores. The login menu item is not shown anyway if the user is already logged in. The session token comes through the usage of Redux.

## PART III: The Presentation Layer (React/HTML)

Once the user types in their user name and password and then click "Login", the `handleLogin()` method makes the call to the backend to get a token. It sets the email and password in its HTTP POST request. If successful, the message RECEIVE_TOKEN_SUCCESS is sent through Redux with a `dispatch()` call. This sets the session token for the rest of the application to see. Then, there is a change made to the browser location to go to the news page view.

The `handleRegistration()` method makes the call to the backend to create a user account in the data layer. The other methods you see are for the binding of the data to the state and the opening and closing of the modal registration dialog via the state property for that.

Here is the rest of the code, minus the render code that you have already seen:

```
import React, { Component } from 'react';
import React, { Component } from 'react';
import PropTypes from 'prop-types';
import { Checkbox, Button, Modal, Glyphicon } from 'react-bootstrap';
import { connect } from 'react-redux'
import superagent from 'superagent';
import noCache from 'superagent-no-cache';
import { FieldGroup } from '../utils/utils';
import '../App.css';

class LoginView extends Component {
  constructor(props) {
    super(props);

    this.state = {
      name: "",
      email: "",
      password: "",
      rememberMe: false,
      showModal: false
    };
  }

  handleRegister = (event) => {
    const { dispatch } = this.props
    event.preventDefault();
    return fetch('/api/users', {
      method: 'POST',
      headers: new Headers({
        'Content-Type': 'application/json'
      }),
      cache: 'default', // no-store or no-cache ro default?
      body: JSON.stringify({
        displayName: this.state.name,
        email: this.state.email,
        password: this.state.password
      })
```

## Chapter 17: NewsWatcher App Development with React

```
    })
    .then(r => r.json().then(json => ({ok:r.ok, status:r.status, json })))
    .then(response => {
      if (!response.ok || response.status !== 201) {
        throw new Error(response.json.message);
      }
      dispatch({ type: 'MSG_DISPLAY', msg: "Registered" });
      this.setState({ showModal: false });
    })
    .catch(error => {
      dispatch({ type: 'MSG_DISPLAY', msg: `Registration failure:
${error.message}` });
    });
  }

  handleLogin = (event) => {
    const { dispatch } = this.props
    event.preventDefault();
    return fetch('/api/sessions', {
      method: 'POST',
      headers: new Headers({
        'Content-Type': 'application/json'
      }),
      cache: 'default', // no-store or no-cache ro default?
      body: JSON.stringify({
        email: this.state.email,
        password: this.state.password
      })
    })
    .then(r => r.json().then(json => ({ok:r.ok, status:r.status, json})))
    .then(response => {
      if (!response.ok || response.status !== 201) {
        throw new Error(response.json.message);
      }
      // Set the token in client side storage if the user desires
      if (this.state.rememberMe) {
        var xfer = {
          token: response.json.token,
          displayName: response.json.displayName,
          userId: response.json.userId
        };
        window.localStorage.setItem("userToken", JSON.stringify(xfer));
      } else {
        window.localStorage.removeItem("userToken");
      }
      dispatch({ type: 'RECEIVE_TOKEN_SUCCESS', msg: `Signed in as
${response.json.displayName}`, session: response.json });
      window.location.hash = "#news";
    })
    .catch(error => {
      dispatch({ type: 'MSG_DISPLAY', msg: `Sign in failed:
${error.message}` });
    });
```

## PART III: The Presentation Layer (React/HTML)

```
  }

  handleNameChange = (event) => {
    this.setState({ name: event.target.value });
  }

  handleEmailChange = (event) => {
    this.setState({ email: event.target.value });
  }

  handlePasswordChange = (event) => {
    this.setState({ password: event.target.value });
  }

  handleCheckboxChange = (event) => {
    this.setState({ remeberMe: event.target.checked });
  }

  handleOpenRegModal = (event) => {
    this.setState({ showModal: true });
  }

  handleCloseRegModal = (event) => {
    this.setState({ showModal: false });
  }
```

**...render() and _renderRegisterModal() left out...**

```
LoginView.propTypes = {
  dispatch: PropTypes.func.isRequired,
  session: PropTypes.object
};

const mapStateToProps = state => {
  return {session: state.app.session}
}

export default connect(mapStateToProps)(LoginView)
```

Chapter 17: NewsWatcher App Development with React

# 17.5 Displaying the News
(src/views/newsview.js and src/views/homenewsview.js)

This NewsView component displays the filtered news page. The UI looks as follows:

*Figure 100 - NewsWatcher news page*

There is a dropdown list that displays the list of news filters to select between. The JavaScript `map()` function is used to go through the array of filters and populate that dropdown. Each one is given text from the `filter.name` property. There is another use of the newsFilter array with the `map()` function to render each of the news stories for the selected filter. A link is provided for each story, along with the image and URL to click and open. There is a lot of usage of Bootstrap components in this code. Here is the render method:

```
render() {
  if (this.props.isLoading) {
    return (
      <h1>Loading news...</h1>
    );
  }
  return (
    <div>
      <h1>News</h1 >
      <FormGroup controlId="formControlsSelect">
        <FormControl bsSize="lg" componentClass="select"
                     placeholder="select"
                     onChange={this.handleChangeFilter}
                     value={this.state.selectedIdx}>
```

319

## PART III: The Presentation Layer (React/HTML)

```
            {this.props.newsFilters.map((filter, idx) =>
              <option key={idx} value={idx}>
                <strong>{filter.name}</strong>
              </option>
            )}
          </FormControl>
        </FormGroup>
        <hr />
        <Media.List>

{this.props.newsFilters[this.state.selectedIdx].newsStories.map((story,
idx) =>
            <Media.ListItem key={idx}>
              <Media.Left>
                <a href={story.link} target="_blank">
                  <img alt="" className="media-object" src={story.imageUrl}
/>
                </a>
              </Media.Left>
              <Media.Body>
                <Media.Heading><b>{story.title}</b></Media.Heading>
                <p>{story.contentSnippet}</p>
                {story.source} <span>{story.hours}</span>
                <Media.Body>
                    <a style={{ cursor: 'pointer' }} onClick={(event) =>
this.handleShareStory(idx, event)}>Share</a>
                </Media.Body>
              </Media.Body>
            </Media.ListItem>
        )}
        <Media.ListItem key={999}>
          <Media.Left>
            <a href="http://developer.nytimes.com" target="_blank"
               rel="noopener noreferrer">
              <img alt="" src="poweredby_nytimes_30b.png" />
            </a>
          </Media.Left>
          <Media.Body>
            <Media.Heading><b>Data provided by The New York
Times</b></Media.Heading>
          </Media.Body>
        </Media.ListItem>
      </Media.List>
    </div>
  );
}
```

There is also the ability to click on a link to Share stories.

## Chapter 17: NewsWatcher App Development with React

**Component supporting code**

When the component is opened, code is run to do the fetching of the news stories. This is done in the `componentDidMount()` method. You can also see at the top of the code how the constructor sets up the state for the selected news filter. The isLoading property comes from Redux and is used for a UI indication that the news stories are being fetched. The message goes away once the data is available. The `handleShareStory()` method is provided so that a story can be sent to the backend to be put in a common location for all users to see and comment on.

The `toHours()` is this little helper function that is being used to format the text to display how old a news story is. Here is the code for the newsview.js, with the render method left out, as that was already shown:

```
import React, { Component } from 'react';
import PropTypes from 'prop-types';
import { FormGroup, FormControl, Media } from 'react-bootstrap';
import { connect } from 'react-redux'
import superagent from 'superagent';
import noCache from 'superagent-no-cache';
import { toHours } from '../utils/utils';
import '../App.css';

class NewsView extends Component {
  constructor(props) {
    super(props);

    this.state = {
      selectedIdx: 0
    };
  }

  componentDidMount() {
    if (!this.props.session) {
      return window.location.hash = "";
    }

    const { dispatch } = this.props
    dispatch({ type: 'REQUEST_NEWS' });
    fetch(`/api/users/${this.props.session.userId}`, {
      method: 'GET',
      headers: new Headers({
        'x-auth': this.props.session.token
      }),
      cache: 'default' // no-store or no-cache?
    })
    .then(r => r.json().then(json => ({ok: r.ok, status: r.status, json})))
    .then(response => {
      if (!response.ok || response.status !== 200) {
        throw new Error(response.json.message);
      }
```

321

# PART III: The Presentation Layer (React/HTML)

```
      for (var i = 0; i < response.json.newsFilters.length; i++) {
        for (var j = 0; j <
              response.json.newsFilters[i].newsStories.length; j++)
        {
           response.json.newsFilters[i].newsStories[j].hours =
toHours(response.json.newsFilters[i].newsStories[j].date);
        }
      }
      dispatch({ type: 'RECEIVE_NEWS_SUCCESS', newsFilters:
response.json.newsFilters });
      dispatch({ type: 'MSG_DISPLAY', msg: "News fetched" });
    })
    .catch(error => {
      dispatch({ type: 'MSG_DISPLAY', msg: `News fetch failed:
${error.message}` });
    });
  }

  handleChangeFilter = (event) => {
    this.setState({ selectedIdx: parseInt(event.target.value, 10) });
  }

  handleShareStory = (index, event) => {
    const { dispatch } = this.props
    event.preventDefault();
    fetch('/api/sharednews', {
      method: 'POST',
      headers: new Headers({
        'x-auth': this.props.session.token,
        'Content-Type': 'application/json'
      }),
      cache: 'default', // no-store or no-cache ro default?
      body: JSON.stringify(
          this.props.newsFilters[this.state.selectedIdx].newsStories[index])
    })
    .then(r => r.json().then(json => ({ok: r.ok, status: r.status, json}))) 
    .then(response => {
      if (!response.ok || response.status !== 201) {
        throw new Error(response.json.message);
      }
      dispatch({ type: 'MSG_DISPLAY', msg: "Story shared" });
    })
    .catch(error => {
      dispatch({ type: 'MSG_DISPLAY',
                 msg: `Share of story failed: ${error.message}` });
    });
  }

  render() {
    ...TAKEN OUT...ALREADY SHOWN...
  }
}
```

```
NewsView.propTypes = {
  dispatch: PropTypes.func.isRequired
};

const mapStateToProps = state => {
  return {
    session: state.app.session,
    newsFilters: state.news.newsFilters,
    isLoading: state.news.isLoading
  }
}

export default connect(mapStateToProps)(NewsView)
```

The home news story page is just a stripped down version of the code seen above, so there is no need to explain it. That is found in the HomeNewsView component.

# 17.6 Shared News Page
### (src/views/sharednewsview.js)

The shared news story view has the same type of news listing capability you have seen before. Here is the render code:

```
render() {
  if (this.props.isLoading) {
    return (<h1>Loading shared news...</h1>);
  }
  return (
    <div>
      <h1>Shared News</h1>
      <Media.List>
        {this.props.news.map((sharedStory, idx) =>
          <Media.ListItem key={idx}>
            <Media.Left>
              <a href={sharedStory.story.link} target="_blank">
                <img alt="" className="media-object"
                    src={sharedStory.story.imageUrl} />
              </a>
            </Media.Left>
            <Media.Body>
              <Media.Heading><b>{sharedStory.story.title}</b></Media.Heading>
                <p>{sharedStory.story.contentSnippet}</p>
                {sharedStory.story.source} -
                 <span>{sharedStory.story.hours}</span>
                <a style={{ cursor: 'pointer' }} onClick={(event) =>
                    this.handleOpenModal(idx, event)}> Comments</a>
            </Media.Body>
          </Media.ListItem>
        )}
```

## PART III: The Presentation Layer (React/HTML)

```
        <Media.ListItem key={999}>
          <Media.Left>
            <a href=http://developer.nytimes.com
               target="_blank" rel="noopener noreferrer">
              <img alt="" src="poweredby_nytimes_30b.png" />
            </a>
          </Media.Left>
          <Media.Body>
            <Media.Heading><b>Data provided by The New York
Times</b></Media.Heading>
          </Media.Body>
        </Media.ListItem>
      </Media.List>
      {this.props.news.length > 0 &&
        <Modal show={this.state.showModal} onHide={this.handleCloseModal}>
          <Modal.Header closeButton>
            <Modal.Title>Add Comment</Modal.Title>
          </Modal.Header>
          <Modal.Body>
            <form onSubmit={this.handleAddComment}>
              <FormGroup controlId="commentList">
                <ControlLabel><Glyphicon glyph="user" />
                Comments</ControlLabel>
                <ul style={{ height: '10em', overflow: 'auto',
                         'overflow-x': 'hidden' }}>
                    {this.props.news[this.state.selectedStoryIdx].
                     comments.map(comment =>
                        <li>
                          <div>
                            <p>'{comment.comment}' - {comment.displayName}
                            </p>
                          </div>
                        </li>
                    )}
                </ul>
              </FormGroup>
                {this.props.news[this.state.selectedStoryIdx].
                 comments.length < 30 &&
                  <div>
                    <FieldGroup
                      id="formControlsComment"
                      type="text"
                      glyph="user"
                      label="Comment"
                      placeholder="Enter your comment"
                      onChange={this.handleCommentChange}
                    />
                    <Button disabled={this.state.comment.length === 0}
                            bsStyle="success" bsSize="lg"
                            block type="submit">
                      <Glyphicon glyph="off" /> Add
                    </Button>
                  </div>
```

## Chapter 17: NewsWatcher App Development with React

```
          }
        </form>
      </Modal.Body>
      <Modal.Footer>
        <Button bsStyle="danger" bsSize="default"
               onClick={this.handleCloseModal}>
          <Glyphicon glyph="remove" /> Close
        </Button>
      </Modal.Footer>
    </Modal>
  }
  </div>
  );
}
```

There is a capability to comment on each story. That is done through a modal dialog similar to the one used for user registration. Here is the image of the viewing of comments:

*Figure 101 - NewsWatcher add comment UI*

### Component supporting code
This code is very similar to the NewsView component. The difference is in the code for adding a new comment. Here is the code:

```
import React, { Component } from 'react';
import PropTypes from 'prop-types';
import { FormGroup, ControlLabel, Button, Modal, Glyphicon, Media } from 'react-bootstrap';
import { connect } from 'react-redux';
import superagent from 'superagent';
```

325

# PART III: The Presentation Layer (React/HTML)

```javascript
import noCache from 'superagent-no-cache';
import { FieldGroup, toHours } from '../utils/utils';
import '../App.css';

class SharedNewsView extends Component {
  constructor(props) {
    super(props);

    this.state = {
      comment: "",
      selectedStoryIdx: 0
    };
  }

  componentDidMount() {
    if (!this.props.session) {
      return window.location.hash = "";
    }

    const { dispatch } = this.props
    dispatch({ type: 'REQUEST_SHAREDNEWS' });
    fetch('/api/sharednews', {
      method: 'GET',
      headers: new Headers({
        'x-auth': this.props.session.token
      }),
      cache: 'default' // no-store or no-cache?
    })
    .then(r => r.json().then(json => ({ok: r.ok, status: r.status, json})))
    .then(response => {
      if (!response.ok || response.status !== 200) {
        throw new Error(response.json.message);
      }
      for (var i = 0; i < response.json.length; i++) {
        response.json[i].story.hours =
          toHours(response.json[i].story.date);
      }
      dispatch({type: 'RECEIVE_SHAREDNEWS_SUCCESS', news: response.json});
      dispatch({ type: 'MSG_DISPLAY', msg: "Shared News fetched" });
    })
    .catch(error => {
      dispatch({ type: 'MSG_DISPLAY',
                 msg: `Shared News fetch failed: ${error.message}` });
    });
  }

  handleOpenModal = (index, event) => {
    this.setState({ selectedStoryIdx: index, showModal: true });
  }

  handleCloseModal = (event) => {
    this.setState({ showModal: false });
  }
```

## Chapter 17: NewsWatcher App Development with React

```
  handleAddComment = (event) => {
    const { dispatch } = this.props
    event.preventDefault();
fetch(`/api/sharednews/${this.props.news[this.state.selectedStoryIdx].story
.storyID}/Comments`, {
      method: 'POST',
      headers: new Headers({
        'x-auth': this.props.session.token,
        'Content-Type': 'application/json'
      }),
      cache: 'default', // no-store or no-cache ro default?
      body: JSON.stringify({ comment: this.state.comment })
    })
    .then(r => r.json().then(json => ({ok: r.ok, status: r.status, json})))
    .then(response => {
      if (!response.ok || response.status !== 201) {
        throw new Error(response.json.message);
      }
      var storyIdx = this.state.selectedStoryIdx;
      dispatch({ type: 'ADD_COMMENT_SUCCESS', comment: this.state.comment,
                 displayName: this.props.session.displayName,
                 storyIdx: storyIdx });
      this.setState({ showModal: false, comment: "" });
      dispatch({ type: 'MSG_DISPLAY', msg: "Comment added" });
    })
    .catch(error => {
      dispatch({ type: 'MSG_DISPLAY',
                 msg: `Comment add failed: ${error.message}` });
    });
  }

  handleCommentChange = (event) => {
    this.setState({ comment: event.target.value });
  }

  render() {
    ...TAKEN OUT...ALREADY SHOWN...
  }
}

SharedNewsView.propTypes = {
  dispatch: PropTypes.func.isRequired
};

const mapStateToProps = state => {
  return {
    session: state.app.session,
    news: state.sharednews.news,
    isLoading: state.sharednews.isLoading
  }
}

export default connect(mapStateToProps)(SharedNewsView)
```

PART III: The Presentation Layer (React/HTML)

# 17.7 Profile Page (src/views/profileview.js)

The profile page allows the user to create one or more news filters. Each filter has a title and a list of keywords. You have the same type of button drop-down you have seen before. Then you have the form and the three buttons to save, delete, and create a new filter. The image and HTML are as follows:

*Figure 102 - NewsWatcher News Filter dialog*

There is a link to allow the user to delete their account. The HTML provides a modal dialog like you have used before. Here is the image:

*Figure 103 - NewsWatcher unregister dialog*

328

```
render() {
  if (this.props.isLoading) {
    return (
      <h1>Loading profile...</h1>
    );
  }
  return (
    <div>
      <h1>Profile: News Filters</h1>
      <FormGroup controlId="formControlsSelect">
        <FormControl bsSize="lg" componentClass="select"
                     placeholder="select"
                     onChange={this.handleChangeFilter}
                     value={this.state.selectedIdx}>
          {this.props.user.newsFilters.map((filter, idx) =>
            <option key={idx}
                    value={idx}><strong>{filter.name}</strong>
            </option>
          )}
        </FormControl>
      </FormGroup>
      <hr />
      <form>
        <FieldGroup
          id="formControlsName"
          type="text"
          label="Name"
          placeholder="NewFilter"
          onChange={this.handleNameChange}
            value={this.props.user.
              newsFilters[this.state.selectedIdx].name}
        />
        <FieldGroup
          id="formControlsKeywords"
          type="text"
          label="Keywords"
          placeholder="Keywords"
          onChange={this.handleKeywordsChange}
            value={this.props.user.
              newsFilters[this.state.selectedIdx].keywordsStr}
        />
        <div class="btn-group btn-group-justified" role="group"
             aria-label="...">
          <ButtonToolbar>
            <Button bsStyle="primary" bsSize="default"
                    onClick={this.handleAdd}><Glyphicon glyph="plus" /> Add
            </Button>
            <Button bsStyle="primary" bsSize="default"
              onClick={this.handleDelete}><Glyphicon glyph="trash"/> Delete
            </Button>
            <Button bsStyle="primary" bsSize="default"
              onClick={this.handleSave}><Glyphicon glyph="save" /> Save
            </Button>
```

```
            </ButtonToolbar>
          </div>
        </form>
        <hr />
        <p>No longer have a need for NewsWatcher? <a id="deleteLink"
            style={{ cursor: 'pointer' }}
            onClick={this.handleOpenModal}>Delete your NewsWatcher Account</a>
        </p>
        <Modal show={this.state.showModal} onHide={this.handleCloseModal}>
          <Modal.Header closeButton>
            <Modal.Title>Un-Register</Modal.Title>
          </Modal.Header>
          <Modal.Body>
            <form onSubmit={this.handleUnRegister}>
              <Checkbox checked={this.state.deleteOK}
                        onChange={this.handleCheckboxChange}>
              Check if you are sure you want to delete your NewsWatcher account
              </Checkbox>
              <Button disabled={!this.state.deleteOK} bsStyle="success"
                      bsSize="lg" block type="submit">
                <Glyphicon glyph="off" /> Delete NewsWatcher Account
              </Button>
            </form>
          </Modal.Body>
          <Modal.Footer>
            <Button bsStyle="danger" bsSize="default"
              onClick={this.handleCloseModal}><Glyphicon glyph="remove" />
                Cancel
            </Button>
          </Modal.Footer>
        </Modal>
      </div>
    );
}
```

## Component supporting code
All the functions necessary to provide the functionality behind the button clicks are made available and should look very similar to code you have seen before.

```
import React, { Component } from 'react';
import PropTypes from 'prop-types';
import { FormGroup, FormControl, Checkbox, Button, Modal, Glyphicon,
ButtonToolbar } from 'react-bootstrap';
import { connect } from 'react-redux';
import superagent from 'superagent';
import noCache from 'superagent-no-cache';
import { FieldGroup } from '../utils/utils';
import '../App.css';

class ProfileView extends Component {
  constructor(props) {
    super(props);
```

## Chapter 17: NewsWatcher App Development with React

```
    this.state = {
      deleteOK: false,
      selectedIdx: 0,
    };
  }

  componentDidMount() {
    if (!this.props.session) {
      return window.location.hash = "";
    }

    const { dispatch } = this.props
    dispatch({ type: 'REQUEST_PROFILE' });
    fetch(`/api/users/${this.props.session.userId}`, {
      method: 'GET',
      headers: new Headers({
        'x-auth': this.props.session.token
      }),
      cache: 'default' // no-store or no-cache?
    })
    .then(r => r.json().then(json => ({ok: r.ok, status: r.status, json})))
    .then(response => {
      if (!response.ok || response.status !== 200) {
        throw new Error(response.json.message);
      }
      for (var i = 0; i < response.json.newsFilters.length; i++) {
        response.json.newsFilters[i].keywordsStr =
          response.json.newsFilters[i].keyWords.join(',');
      }
      dispatch({ type: 'RECEIVE_PROFILE_SUCCESS', user: response.json });
      dispatch({ type: 'MSG_DISPLAY', msg: "Profile fetched" });
    })
    .catch(error => {
      dispatch({ type: 'MSG_DISPLAY',
        msg: `Profile fetch failed: ${error.message}` });
    });
  }


  handleUnRegister = (event) => {
    const { dispatch } = this.props
    event.preventDefault();
    fetch(`/api/users/${this.props.session.userId}`, {
      method: 'DELETE',
      headers: new Headers({
        'x-auth': this.props.session.token
      }),
      cache: 'default' // no-store or no-cache?
    })
    .then(r => r.json().then(json => ({ok: r.ok, status: r.status, json})))
    .then(response => {
      if (!response.ok || response.status !== 200) {
```

```
          throw new Error(response.json.message);
        }
        this.props.appLogoutCB();
        dispatch({ type: 'MSG_DISPLAY', msg: "Account deleted" });
      })
      .catch(error => {
        dispatch({ type: 'MSG_DISPLAY',
                   msg: `Account delete failed: ${error.message}` });
      });
  }

  handleNameChange = (event) => {
    this.props.dispatch({ type: 'ALTER_FILTER_NAME',
        filterIdx: this.state.selectedIdx, value: event.target.value });
  }

  handleKeywordsChange = (event) => {
    this.props.dispatch({ type: 'ALTER_FILTER_KEYWORDS',
        filterIdx: this.state.selectedIdx, value: event.target.value });
  }

  handleOpenModal = (event) => {
    this.setState({ showModal: true });
  }

  handleCloseModal = (event) => {
    this.setState({ showModal: false });
  }

  handleChangeFilter = (event) => {
    this.setState({ selectedIdx: parseInt(event.target.value, 10) });
  }

  handleAdd = (event) => {
    const { dispatch } = this.props
    event.preventDefault();
    if (this.props.user.newsFilters.length === 5) {
      dispatch({ type: 'MSG_DISPLAY',
          msg: "No more newsFilters allowed" });
    } else {
      var len = this.props.user.newsFilters.length;
      dispatch({ type: 'ADD_FILTER' });
      this.setState({ selectedIdx: len });
    }
  }

  handleDelete = (event) => {
    event.preventDefault();
    this.props.dispatch({ type: 'DELETE_FILTER',
        selectedIdx: this.state.selectedIdx });
    this.setState({ selectedIdx: 0 });
  }
```

```
  handleSave = (event) => {
    const { dispatch } = this.props
    event.preventDefault();
    fetch(`/api/users/${this.props.session.userId}`, {
      method: 'PUT',
      headers: new Headers({
        'x-auth': this.props.session.token,
        'Content-Type': 'application/json'
      }),
      cache: 'default', // no-store or no-cache ro default?
      body: JSON.stringify(this.props.user)
    })
    .then(r => r.json().then(json => ({ok: r.ok, status: r.status, json})))
    .then(response => {
      if (!response.ok || response.status !== 200) {
        throw new Error(response.json.message);
      }
      dispatch({ type: 'MSG_DISPLAY', msg: "Profile saved" });
    })
    .catch(error => {
      dispatch({ type: 'MSG_DISPLAY',
          msg: `Profile save failed: ${error.message}` });
    });
  }

  handleCheckboxChange = (event) => {
    this.setState({ deleteOK: event.target.checked });
  }

  render() {
    ...TAKEN OUT...ALREADY SHOWN...
  }
}

ProfileView.propTypes = {
  appLogoutCB: PropTypes.func.isRequired,
  dispatch: PropTypes.func.isRequired
};

const mapStateToProps = state => {
  return {
    session: state.app.session,
    user: state.profile.user,
    isLoading: state.profile.isLoading
  }
}

export default connect(mapStateToProps)(ProfileView)
```

You can see that there is code there to limit the number of filters to five. What do you suppose would happen if the user used the browser developer tools to mess with that code and then added thousands of filters? It could create a very large documents in MongoDB. Be aware

## PART III: The Presentation Layer (React/HTML)

that you cannot ever rely on your UI side bounds-checking code to do the right thing, as it is subject to tampering. Thus, you have to place code in your service layer that will only take the first five filters.

It is also noteworthy, how the function to handle the logging out is not in this code. Instead, the App Component that uses the ProfileView component passes in a function as a property on the props and then the callback is used to get at code in App.js for that. This seemed to be more appropriate for the higher calling code to control that and not need to duplicate code in this case.

# 17.8 Not Found Page (src/views/notfound.js)

The Not Found page is something that would be displayed if the user tried to navigate by changing the URL in the browser to some page that did not exist. You can see this in action if you try and go to https://www.newswatcher2rweb.com/#/blah. Here is that Component code.

```
import React, { Component } from 'react';
import '../App.css';

class NotFound extends Component {
  render() {
    return (
      <div>
        <h3>404 page not found</h3>
        <p>The page you are looking for does not exist.</p>
      </div>
    );
  }
}

export default NotFound;
```

This completes the discussion of each of the files and their corresponding controller code. Everything can be zipped up and deployed to the AWS Elastic Beanstalk environment and used. On Windows, you would select the folders and files as shown before and then right click and select **Send to->Compressed (zipped) folder**.

# Chapter 18: UI Testing of NewsWatcher

With the NewsWatcher application code completed and running, you will want to thoroughly test it with every conceivable scenario. Manually doing this is great for your entertainment, but it will soon grow tedious if you have that as your only way of finding bugs. For example, what happens if you make any changes to your code base? You have the function and load tests of the API, but you still need to run the UI through its paces to exercise the React JavaScript code. Not to fear, there are many great solutions to this problem and I will present this as well as give general tips for debugging the code in Chrome.

There are many different technologies being used in the code and each has techniques for testing it in detail. For example, the React Router usage and Redux code can be independently tested. Each individual component can be tested in ways that instantiate them and verify correct DOM elements are present with the correct attributes and properties. Then you can also try testing frameworks that do an automated run through of the complete application in a browser. You use IDs of each UI element to identify what to click or inspect.

## 18.1 UI Testing with Selenium

There are tools you can use to record your interactions within a browser session and then play those back. Some tools record screen locations of your clicks and then rely on the UI elements to be in that exact same location later for the interaction to work. Other tools understand the DOM and can find elements you specify to click on.

One tool you can use to do automation testing of a UI, is a tool created a while a while ago named Selenium. It was built so long ago that it did not support JavaScript directly. Luckily, someone did the work to write a node.js module that exposes its capabilities.

The first step towards accomplishing your UI automation testing will be to install the "selenium-webdriver" from NPM to have it local in your project. Be aware that this might not be as simple as it sounds. Pay attention to the output window as you install it. It might fail because of dependencies. I had to install the JDK, Python tools, and even a more complete Visual Studio install with some of the C++ tools. I would not have dreamed those would be required. Your experience might be different on a different OS.

You might also check out an NPM module named Protractor. This is something written on top of selenium-webdriver. I decided to go directly through the selenium-webdriver module, so did not use Protractor.

## PART III: The Presentation Layer (React/HTML)

The selenium-webdriver is just an SDK to access the browser DOM and thus you still need some type of testing framework to organize and run your tests. Mocha is perfect for doing this. This is what you would run from the command prompt to get your UI automation tests running.

```
.\node_modules\.bin\mocha --timeout 30000 ui_automation_UAT.js
```

You will populate a Mocha test file with code that uses the selenium-webdriver capabilities. You also need to set up the `require` statements for asserts and for making HTTP request to do some needed cleanup after a test run. Here are the require statements:

```
var assert = require('assert');
var webdriver = require('selenium-webdriver');
var request = require('supertest')('http://localhost:3000);
```

You will use the usual Mocha `describe` and `it` code blocks. You need to first set up some code that runs before any tests in a `before` block. This code initializes Selenium and sets up which browser you want it to use. The `after` block does the cleanup.

Using the driver object, you can get access to any element in the DOM, to inspect it or affect it in some way. For example, with an HTML text control, you can send it keystrokes to enter values as a user would. Buttons can be clicked, etc.

Many times, you do something like a button click and then need to wait for the UI to respond. The driver has a `wait()` function you use to wait for UI elements to be available. For example, if you were navigating to a new page, you would use a wait to block until the page was ready. The `wait()` function takes as a first parameter, what you are waiting for. You can thus wait for an element to be visible. Inside that function, you provide the id that will be used by the webdriver object to locate the HTML element. There is a time limit set for how long it should wait. It will wait up to that time, but if the element appears before that, it will immediately move on. Sometimes you will need to put in a `setTimeout()` call to do the delay. This should be avoided as these add up and add to the total time it takes for a test run to complete.

Here is the specifying of an id in your HTML that you then can use to identify a button by later in your test code.

```
<button id="btnRegister" ng-click="register()">Register</button>
```

As with all Mocha tests, you need to call `done()` when that test is finished and move on to the next test.

# Chapter 18: UI Testing of NewsWatcher

The whole point of putting together a UI automation test with Selenium is to mimic what you would normally do manually and thus have a repeatable test suite that you can run as part of a CI/CD script and save you a lot of time.

There is the concept of a User Acceptance Test or UAT that basically is the user script that you want to follow to prove that the UI can do everything it is supposed to do. Development teams can create a UAT for each code iteration they go through before a deployment can be approved.

You can set up the test to go against your staging, production or local hosted site. Here is a small sample of the code for the UI automation testing that uses Selenium. I have set it up to use Internet Explorer, but you can use any browser you like, even a hidden one. You will get a message if you do not have the driver installed for the browser you want to use. For example, with IE, the file is IEDriverServer.exe. You will see a message on where to get it from. You just unzip it and copy it to a location that is in your path.

## NewsWatcher Selenium tests

This code simply launches the site and then clicks on the login tab and then attempts to log in with a user email that does not exist. The code tests that the correct error message appears. Note how there needs to be delays put into the test code to wait for UI changes to happen.

```
describe('NewsWatcher UI exercising', function () {
  var driver;
  var storyID;
  var token;


  // Runs before all tests in this block
  before(function (done) {
    driver = new webdriver.Builder().withCapabilities(
                webdriver.Capabilities.ie()).build();
    driver.get('http://localhost:3000');
    driver.wait(webdriver.until.elementLocated(
      webdriver.By.id('loginLink')), 10000).then(function (item) {
      done();
    });
  });

  // Runs after all tests in this block
  after(function (done) {
    driver.quit().then(done);
  });

  it('should deny a login with a non-registered email', function (done) {
    driver.findElement(webdriver.By.id('loginLink')).click();
    driver.findElement(webdriver.By.id('formControlsEmail2')).
      sendKeys('T@b.com');
```

```
    driver.findElement(webdriver.By.id('formControlsPassword2')).
      sendKeys('abc123*');
    driver.findElement(webdriver.By.id('btnLogin')).click();
    driver.wait(webdriver.until.elementLocated(
            webdriver.By.id('currentMsgId')), 5000);
    // Wait a few seconds for the UI to update
    setTimeout(function () {
      driver.findElement(webdriver.By.id('currentMsgId')).getText().then(
        function (value)
      {
        assert.equal(value,
          '(Sign in failed: Error: User was not found.)');
        done();
      });
    }, 5000);
  });
});
```

*Note:* If you look at the selenium test code in my GitHub project, you will find most of it commented out. This is because the code was taken from an original Angular application and I decided not to continue with the Selenium testing and instead go with Enzyme. Selenium still has its place and is useful for end-to-end user acceptance testing. You might want to investigate the use of Nightwatch, as a nice wrapper on top of Selenium.

# 18.2 UI Testing with Enzyme

Enzyme is an API that you use to test your React components at the code level. With it you will do component level testing. This is analogous to Unit testing of service layer code. You can isolate a given component and verify that it works as an individual piece of code and then have a greater assurance that when consumed, it will all work together.

You will want to start with tests for components at the lowest levels because it is sometimes best to use code to test code as it is the most efficient means of trying all the combinations of code paths (including error paths). As part of doing this, you will mock data that might be coming from things like HTTP calls. The point is to isolate a component as much as possible.

You can also test components that are at a higher level, such as the ones that consume other components. This means you could even test the complete application from the highest App component and exercise all parts of it.

It is up to you to decide how rigorous you want the Enzyme tests to be. For example, you can test components and verify that when they are instantiated they have the proper state, props and styles. You can go further and code up tests for your Redux reducers. You can also do full testing as if you are interacting with a component with mouse clicks etc.

## Chapter 18: UI Testing of NewsWatcher

***Node:*** *Mocha is often used as the test runner for the Enzyme tests. I have chosen Jest instead (built on top of Jasmine), because that is what is installed by default with the React application that was created. It also has some great features that Mocha does not have, such as built-in code coverage and parallel running of tests. You can't really tell the difference anyway between a mocha test file and a Jest one. They both have 'describe' and 'it' blocks of code. Both also have the ability to use the done() function to do asynchronous testing.*

Enzyme is the API you make use of inside of each test case in code that Jest will run. Enzyme is easy to use. With it, you instantiate a component with either the `shallow()` or `mount()` functions. Shallow does not create a virtual DOM, but is lightweight and just creates the component at that level with no hierarchy underneath it. This works in most cases for your testing purposes.

If you are testing something like a component that is wrapped by Redux, you need to use the `mount()` capability as it will get lifecycle events running. It will have a full virtual DOM, and a redux state store will be used. I will give examples of both the shallow and the mount usage.

## NewsWatcher Enzyme tests with shallow

Think back to the home page UI that had news stories served up. In that React code, you can find a `componentDidMount()` function that goes to the backend to fetch the list of top news stories. The data returned from the fetch call gets placed into the local component state and then the `render()` function updates the UI with the latest data. Here is some test code for verifying all of that, using some mocked data. I placed a lot of comments in the code to help explain it.

```
import React from 'react';
import { shallow } from 'enzyme';
import { createStore } from 'redux';
import reducer from '../reducers';
import HomeNewsView from './homenewsview';

// A helper function to put together an HTTP response for our mocking
const mockResponse = (status, statusText, response) => {
  return new window.Response(response, {
    status: status,
    statusText: statusText,
    headers: {
      'Content-type': 'application/json'
    }
  });
};

describe('<HomeNewsView /> (mocked data)', () => {
  it('news stories are displayed', (done) => {
```

339

## PART III: The Presentation Layer (React/HTML)

```
    // This is the payload that our mocking will return
    const mockData = [{
      contentSnippet: "The launch of a new rocket by Elon Musk's SpaceX.",
      date: 1514911829000,
      hours: "33 hours ago",
      imageUrl: "https://static01.nyt.com/images/blah.jpg",
      link: "https://www.nytimes.com/2018/01/01/science/blah.html",
      source: "Science",
      storyID: "5777",
      title: "Rocket Launches and Trips to the Moon",
    }];

    // We use the actual redux store with our official reducers
    // We replace the JavaScript fetch() function with our own
    // and always return our mock data
    const store = createStore(reducer)
    global.fetch = () => Promise.resolve(mockResponse(200, null,
JSON.stringify(mockData)));

    // Here is the usage of Enzyme to instantiate our component
    const rc = shallow(<HomeNewsView dispatch={store.dispatch} />, {
disableLifecycleMethods: true })
    expect(rc.state().isLoading).toEqual(true);

    // We are doing shallow instantiation, so we need to call by hand
    // the componentDidMount() function and wait on the promise resolve
    // to then do testing against what is expected because of the
    // news story fetch and render with that data
    rc.instance().componentDidMount().then((value) => {
      // All state properties will be updated by now,
      // however, the render may not have happened yet.
      // The update() call is made, so the test of the <h1> element
      // will now have the refreshed value
      rc.update();
      // Verify some of the state that was set
      expect(rc.state().isLoading).toEqual(false);
      // verify some actual elements that were rendered
      expect(rc.find('h1').text()).toEqual('Home Page News');
      const listNews = rc.state().news;
      expect(listNews.length).toEqual(1);
      expect(listNews[0].title).toEqual('Rocket Launches and Trips to the
Moon');
      expect(rc.find('b').first().text()).toEqual('Rocket Launches and
Trips to the Moon');
      // Verify the Redux state was set and there is
      // a successful message on fetch completion.
      expect(store.getState().app.currentMsg).toEqual('Home Page news
fetched');
      done();
    })
  });
});
```

## Chapter 18: UI Testing of NewsWatcher

You can see the use of `shallow()` and then the further usage of the object returned with functions like `find()` and `state()`. These allow you to interrogate the instance for what is expected.

The `expect()` function of Jest is used for the validations. The component returned from the shallow usage can be used to do things like find sub-elements and inspect them. You can see where we find an `h1` element and verify the text of it. You can also inspect the state properties of the component. This is done to verify what news stories were fetched.

Another test is done that inspects the Redux store with a `getState()` call. You can look at anything in the store as you see is happening in the verification of what the `app.currentMsg` is set to.

Since we are doing shallow rendering, the `componentDidMount()` function will not be called as part of the lifecycle operations. That is why you see it being called in test code. The call to `componentDidMount()` is asynchronous, so we needed to make use of the `done()` function to tell Jest that this test is complete.

*Note: You might be tempted to use a setTimeout() in test code and delay for a few seconds until the UI has rendered. This actually will not work in Jest, and is a bad idea anyway. If you had hundreds of tests that each had delays, you would increase the time it takes to run your tests and could not really be guaranteed any race conditions would work out.*

The actual `componentDidMount()` component code gets called, so it will do things like make the fetch call and then try and also set the state in Redux with a dispatch call. Since we want to control the actual returned response, we need to mock all of this up and override the fetch function. We also set up Redux here, because that is normally done in the App component, and we are working at a level below that.

*Note: An alternative to calling the componentDidMount() would be for you to mimic the calls that are happening to the Redux actions. You could place calls to store.dispatch() and then do some expect() calls. You could also set the state or props directly on components and then force an update and then do the expect tests.*

## NewsWatcher Enzyme tests with mount

The code to test the login must be a bit different. This is because the component is needing Redux. In this case, we need to use the `mount()` capability of Enzyme. We have to use `mount()` because we want to use the `connected` component. It has the functionality to use the redux dispatch to update props and we can verify that the store state is correctly set.

Here is some code that mounts the component and then tests that a login was successful.

```
it('User can log in', (done) => {
  const mockData = {
    displayName: "Buzz",
    userId: "1234",
    token: "zzz",
    msg: "Authorized"
  };

  const store = createStore(reducer)
  global.fetch = () => Promise.resolve(mockResponse(201, null,
                      JSON.stringify(mockData)));

  // Need to mock the local storage call as well
  const localStorageMock = {
    setItem: () => { },
    removeItem: () => { }
  };
  global.localStorage = localStorageMock

  const wrapper = mount(<ConnectedLoginView store={store} />)
  let rc = wrapper.find(LoginView);
  expect(rc.props().session).toEqual(null);

  rc.instance().handleLogin({ preventDefault() { } }).then((value) => {
    expect(store.getState().app.session.displayName).toEqual('Buzz');
    expect(store.getState().app.currentMsg).toEqual('Signed in as Buzz');
    done();
  })
});
```

We override fetch again as before. Notice how we also needed to provide a stubbed out `preventDefault()` call to not do anything, as `handleLogin()` is expecting that function to exist.

**Note:** *We could have caused a click to have happened on the login button and the code would have proceeded to execute that. The problem is that we would not know when that finished, so we can simply call the function ourselves and wait for the resolve and do what testing we want after that.*

You will notice we don't even fill in the email and password, as our fetch mocking does not pay attention to anything passed in and returns what we want, under our control.

## UI element interaction

Here is some code that does some manipulation of the actual elements like the text control to hold the email and the checkbox control. You can use shallow or mount and accomplish the same thing. What we are really checking here is the binding of each control by looking at the state properties that should be set as a result of the interactions.

## Chapter 18: UI Testing of NewsWatcher

```
it('User can change remember me checkbox and enter an email', () => {
  const rc = shallow(<LoginView dispatch={() => {}} session={null} />, { disableLifecycleMethods: true })

  expect(rc.state().remeberMe).toEqual(false);
  rc.instance().handleCheckboxChange({ target: {checked: true } });
  expect(rc.state().remeberMe).toEqual(true);

  rc.find('#formControlsEmail2').last().simulate('change',{ target: { value: "abc@def.com" } })
  expect(rc.state().email).toEqual("abc@def.com");
});
```

In the case of a checkbox, there is not a way to click it or cause a change to it to have it run the `onChange()` handler so we just call the `handleCheckboxChange()` method ourselves and fake what would be passed in for the event parameter.

For the email text input element, we can simulate the text entry. In this case, we can get an `onChange()` to happen automatically for us.

# Getting set up to use Enzyme and running tests

Before using Enzyme, you need to install the npm packages as follows:

```
npm install --save enzyme enzyme-adapter-react-16 react-test-renderer
```

Then you also need to create a file in the src directory named setupTests.js with the following contents:

```
// src/setupTests.js
import { configure } from 'enzyme';
import Adapter from 'enzyme-adapter-react-16';

configure({ adapter: new Adapter() });
```

To run the tests you just run the following:

```
npm run test-react
```

You can alter the test script in the package.json file to add the running of the Enzyme tests and also add a flag to tell it to include code coverage numbers.

```
"test": "mocha --timeout 30000 test/functional_api_crud.js && react-scripts test --coverage --env=jsdom",
```

343

PART III: The Presentation Layer (React/HTML)

If you do some searching online and you will find all kinds of creative ways to test with Jest and Enzyme. For example, I did not mention the ability of Enzyme to compare DOM snapshots from one test run to another.

Good tests code can sometimes be as difficult or even more difficult than the actual product code. I have worked on several projects where there was twice as much test code written as production code.

Code coverage numbers look as follows:

```
:/c/Users/eljam/Documents/GitHub/NewsWatcher2RWeb
Ran all test suites.
-----------------|---------|----------|---------|---------|-------------------
File             | % Stmts | % Branch | % Funcs | % Lines | Uncovered Lines
-----------------|---------|----------|---------|---------|-------------------
All files        |   27.94 |    24.17 |   23.64 |   29.02 |
 src             |    5.26 |        0 |       0 |    6.45 |
  App.js         |       5 |        0 |       0 |    5.26 | ... 51,74,78,88
  index.js       |       0 |        0 |       0 |       0 | ... 7,8,9,11,13
  setupTests.js  |     100 |      100 |     100 |     100 |
 src/reducers    |   62.16 |    46.15 |     100 |   62.16 |
  app.js         |   85.71 |       80 |     100 |   85.71 |                22
  homenews.js    |   66.67 |       50 |     100 |   66.67 |              9,14
  index.js       |     100 |      100 |     100 |     100 |
  news.js        |   66.67 |       50 |     100 |   66.67 |              9,14
  profile.js     |      40 |       25 |     100 |      40 | ... 22,40,49,58
  sharednews.js  |   57.14 |       40 |     100 |   57.14 |          11,16,22
 src/utils       |   85.71 |    66.67 |     100 |   85.71 |
  utils.js       |   85.71 |    66.67 |     100 |   85.71 |                 9
 src/views       |   23.68 |    20.31 |   20.65 |   23.89 |
  homenewsview.js|     100 |      100 |     100 |     100 |
  loginview.js   |   58.97 |    58.33 |      55 |   59.46 | ... ,96,104,108
  newsview.js    |    5.56 |        0 |       0 |    5.88 | ... ,96,107,133
  notfound.js    |       0 |      100 |       0 |       0 |                 6
  profileview.js |    3.45 |        0 |       0 |    3.64 | ... 143,149,210
  sharednewsview.js|     5 |        0 |       0 |    5.26 | ... 108,134,174
-----------------|---------|----------|---------|---------|-------------------

> newswatcher@0.0.1 posttest C:\Users\eljam\Documents\GitHub\Newswatcher2RWeb
> echo All tests have been run!

All tests have been run!
```

*Figure 104 - F12 debugger console*

## 18.3 Debugging UI Code Issues

To locally debug you server side node.js code is very simple. By using the VS Code editor, you have the capability to run it locally and set breakpoints. React code, however, runs in the browser. Browsers have their own debugging capabilities that you must learn and make use of. For example, with Chrome, there is a selection in the menu under **More tools** to launch **Developer tools**.

Once open, you can click on the **Console** tab and see errors that happen in your code. This image shows a code error for an undefined property:

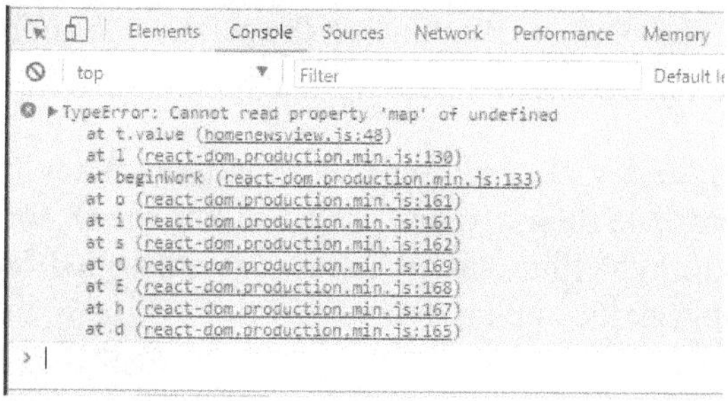

# Chapter 18: UI Testing of NewsWatcher

![debugger console]

*Figure 105 - debugger console in Chrome*

This window is very handy to have open, to see problems you might not otherwise notice. Many times, it is quite easy to go and fix the bugs you find. Other times, you will need to step through the lines of code to tell what is happening. Simply click the **Sources** tab and you can browse your code and set breakpoints in the browser. The experience is typical of any debugger, in that you can inspect variables and set watches on them.

This next image shows how I used the debugger to set a breakpoint and then inspected the response data being returned from an HTTP request. The left pane has a list of files that you can look through. I opened up the JavaScript code for the home news viewing. In the pane where the code is, you can click on a line number and set a breakpoint. You may need to hit the browser refresh button, or click around in your application to make the code run, to hit your breakpoint.

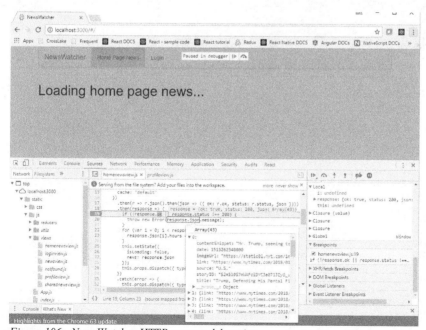

*Figure 106 - NewsWatcher HTTP request debugging*

PART III: The Presentation Layer (React/HTML)

You can have VSCode running the node server and place breakpoints in the server side code and debug back and forth in both environments.

If you have written HTML, you know the capabilities of the browser developer tools and will certainly know how to inspect the DOM elements and their associated CSS. Click the **DOM Explorer** tab to try this out. For example, you might find that an element is not styled or functioning as expected. A quick inspection can usually turn up the issue.

The **Network** tab shows you the HTTP traffic going back and forth similar to what you see using Fiddler or Postman. The **Performance** tab can let you capture the calls going on and then view that in a call graph that shows the time of everything in details. This allows you to check for and fix performance issues. The **Memory** tab lets you set up a run that captures memory snapshots so you can investigate memory leaks in your code.

*Figure 107 - Chrome Network capturing*

# Chapter 19: Server-Side Rendering

The NewsWatcher application has been developed as a purely client-side SPA application that loads and renders on a client device. It is run in the context of a browser, whether that is on a mobile smartphone, tablet, laptop or personal computer. This works quite well.

A SPA like NewsWatcher has great performance and mostly minimizes traffic back and forth to just the data fetching. All the UI is always rendered client-side and is never loaded from the server, once it is all brought over. There is a bit of resistance in the development community to fully sanction SPAs as the way to go. Their objections are based on two points as follows:

1. Performance is better servicing up pages from a server. A server rendered app can collect all the data on the server side and serve it up faster than a SPA. Especially when it comes to the initial page loading, if there is a lot that goes with it.
2. SEO (Search Engine Optimization) is more efficient for server rendered HTML pages because of how Google indexes them.

What some people are advocating, is to move to an SSR (Server Side Rendering) design to address these two issues. There can, of course, be a hybrid solution to give you the best of a SPA and an SSR solution. Let's first take a look at the objections one by one.

As far as the first point, this can be mitigated by implementing code splitting so that not everything is sent to the client at the start. This is where you only send up UI as needed. The code is effectively split up into parts. The initial UI code is sent and renders very fast and then other parts are brought in as needed. This would then allow just as fast of a page load for a SPA as for an SSR site. As you navigate through the UI, pages are brought to the client just in time.

The second point is also debatable in that Google will crawl SPA applications and can even look through JavaScript code while doing web indexing.

*Note: To really do justice to an SSR, you also need to put in meta tags in the HTML as helpers for the web indexers to help with search engine optimization (SEO).*

If you decide to, you can write the site to be completely rendered as an SSR site. You can even have Redux and React Router used on the server. As a compromise, you can make a kind of hybrid implementation, where some of the UI is rendered on the server side and returned as HTML/JavaScript to the client browser and once there, that code can also contain React code and have that code run on both the Browser side as well as on the SPA. This is what a Universal application design pattern is all about – being able to run the same type of code (React Router, Redux, etc.) on the server and on the client.

The hybrid Universal JavaScript solution would certainly address both issues and might be what you want to go towards. This would mean that you have the initial page rendered with SSR, including its Redux store data along with the initial page so that it renders instantly. This gives you the speed for the initial page rendering. The SEO issue would be taken care of for this initial page for Google to index. The page loads immediately and does not have to go back to fetch any additional data until further user interactions are happening.

A Universal hybrid approach is great if you have an initial route page with critical information to be indexed. From there, that page can even function as a SPA, with sub-navigation that goes along with it. You can decide what pages you want served up as SSR routes and then have each of those be SPA pages functioning on their own.

*Note: if you are thinking that a lot of your users will be on mobile devices that have cellphone or Wi-Fi connection anyway, you might as well develop native applications. This means you have a purely native application they download. This is in essence functioning like the SPA does, except that the application is always instantly there on the device. You also get the benefit of writing the native application to work in a mode where it is off-line. You can also investigate what a Progressive Web application is for doing that in the browser.*

You will also want to implement some caching on the server side if you are doing SSR. This is a huge performance advantage. For example, if the home news page is going to stay the same for everyone for a few hours, you render it and serve that up from a cached copy and then update it when a new set of news is ready to go. There are NPM modules you can use that make this seamless and are easy to put into your code.

# 19.1 NewsWatcher and SSR

I decided to experiment with having the initial home news page as a server-side rendered page and then the rest of the application being a client SPA. It turned out to work really well and is an option for those that decide to go this route. It may even turn out that someday `create-react-app` generates code that allows you the option of doing code as SSR.

*Note: You can find my SSR experiment on GitHub. This was a brief attempt and should not be taken as a finished project to pattern anything after. I was able to preserve the original app without doing an eject (https://github.com/eljamaki01/NewsWatcher2RWebSSR). Be aware that I took the code snapshot for this experiment at an early stage of NewsWatcher, so a lot changed between that and the other repository that ended up being the main one for this book.*

## Chapter 19: Server-Side Rendering

What I did, is set things up for the route to the home news page to be rendered with SSR. The rest of the application still works the same as a SPA. The idea is that an initial call comes in for the NewsWatcher site and an HTML page is rendered on the server side for it. This initial page is the home news page. That HTML can also contain a special section that contains all of the actual fetched data for the news stories. This also goes up. You might wonder why in the world that would be going on. The answer has to do with how React works.

React will only render UI that has changed. Thus, an initial page of the home news is rendered and then React also does the work to work as if it is a SPA and takes the data sent up and also tries to create a virtual DOM for the home news from that. It turns out to be identical with what was already being shown in the actual DOM and so it throws it away.

It turns out that the page instantly loads and there is no delay. It all just instantly appears. Other views, such as the user news page will show the loading text while the `componentDidMount()` code chunks away, but the home news page never shows the loading text.

One change that needs to happen is that the server.js file needs to serve up the route for the main `'/'` route that is the UI application itself. That code will end up being in a new file that you create named ssrrender.js. Here is the server.js file code to return the server-rendered main page:

```
process.env.BABEL_ENV = 'production';
process.env.NODE_ENV = 'production';
require('babel-register')({
  ignore: /\/(build|node_modules)\//,
  presets: ['env', 'react-app']
})

const SSRRender = require('./ssrrender');
const SSRRenderRouter = require('./ssrrenderRouter');
app.use('/', SSRRenderRouter)

// Serving up of static content such as HTML, images, CSS files, and
JavaScript files
app.use(express.static(path.join(__dirname, 'build')));
app.use('/', SSRRender)
```

The code to render the static page is going to look familiar. You will even see the use of the Provider component and the Redux store and reducers. It is all there as before. This means that the exact process of rendering a page on the client is now happening on the server. Even the bootstrap style files are there. There are two interesting lines that do string replacements on the HTML template. The first one does the actual replacement of the HTML that was rendered by React. The second one is the one that adds to the Redux state. It is transferred

349

## PART III: The Presentation Layer (React/HTML)

with the news story list up to the client to be used there. This is because the `ComponenbtDidMount()` is not used to retrieve it anymore. Here is all of the code:

```
// ssrrender.js
const path = require('path')
const fs = require('fs')
const config = require('./config');

const React = require('react')
const { renderToString } = require('react-dom/server')
const { StaticRouter } = require('react-router-dom')

const { createStore } = require('redux')
const { Provider } = require('react-redux')
const { default: App } = require('./src/App')
const { default: reducer } = require('./src/reducers')
import { toHours } from './src/utils/utils';

var express = require('express');
import 'bootstrap/dist/css/bootstrap.css';
import 'bootstrap/dist/css/bootstrap-theme.css';
var router = express.Router();

// Return all the Home Page news stories. Verify we have a logged in user.
module.exports = function handleSSR(req, res, next) {
  req.db.collection.findOne({ _id: config.GLOBAL_STORIES_ID },
                            { homeNewsStories: 1 },
                            function (err, doc)
{
    if (err) return next(err);

    // Populate an initial state
    for (var i = 0; i < doc.homeNewsStories.length; i++) {
      doc.homeNewsStories[i].hours = toHours(doc.homeNewsStories[i].date);
    }
    let preloadedState = { homenews: { isLoading: false,
                                        news: doc.homeNewsStories } }

    // Create a new Redux store instance
    const store = createStore(reducer, preloadedState)
    const context = {}
    const html = renderToString(
      <Provider store={store}>
        <StaticRouter
          location={req.url}
          context={context}
        >
          <App />
        </StaticRouter>
      </Provider>
    )
```

## Chapter 19: Server-Side Rendering

```
    // Grab the initial state from our Redux store
    const finalState = store.getState()

    // Send the rendered page back to the client
    const filePath = path.resolve(__dirname, 'build', 'index.html')
    fs.readFile(filePath, 'utf8', (err, htmlData) => {
      if (err) {
        console.error('read err', err)
        return res.status(404).end()
      }

      // We're good, so send the response
      let replace1 = htmlData.replace('{{SSR}}', html)
      let replace2 = replace1.replace(`console.log("REPLACE")`,
        `window.__PRELOADED_STATE__ =
${JSON.stringify(finalState).replace(/</g, '\\u003c')}`)
      res.send(replace2)
    })
  });
};
```

There is one minor tweak to the homenewsview.js file. You simply comment out the `componentDidMount()` function, because it already has the news stories in the initial load and that is placed in the redux store to be used.

The file index.js also has a minor tweak where it uses a special window object property `window.__PRELOADED_STATE__` property that has the news stories and loads that into the Redux store on the client side. You see that on the server-side code in ssrrender.js that the store is also preloaded with that data. You can look there and you will find the actual call to the database. There is no HTTP call, as on the server side we have the database connection to use directly.

*Note*: *Server-side rendering is not for everyone. Whatever you do, you would have to weigh the costs versus the benefit to help you make the right decision. It can be tricky to get right and can certainly make your code more complicated to understand going forward.*

# Chapter 20: Native Mobile Application development with React Native

The existing NewsWatcher React SPA client code can be taken and transformed into code that runs as a native application on a mobile device. The React Web app was built to be responsive, meaning adapt to the screen size that it is running on. It was built using Bootstrap to give it such things as a collapsible menu when running on a mobile phone in the browser. This still does not compare to what performance and UI capabilities can be accomplished with a truly native application.

One benefit of Native applications is that they can be designed to run in an off-line mode when no data or WIFI connection is available. Sure, there is the possibility of doing a progressive web application, but why not just stick with the real thing and go completely native? Native applications have much better performance than a browser-based application running on a phone. The other big differentiator is that Native applications can access all the hardware features of the phone itself, such as the camera, contacts, gyroscope and many other features.

You simply do a bit of tweaking to the code in the render functions and then use a tool to create a native application file that is appropriate for either Google Play or for the Apple App store.

A React Native application runs the low-level mobile platform code through a JavaScript "bridge" layer to interact with the phone for accessing all its functionality and presenting a UI. There is a JavaScript engine interface that exists on both iOS and Android. This is a low-level layer that then interacts with the OS for the phone to do the same things that other platform languages on the phones do. You may have heard of Objective-C, Java, Xcode or Swift.

React web sites write to the HTML DOM, as that is the layer for all browsers, but for phones, it is the JavaScript engine layer as the point of interaction. This means you get the most performant applications possible, and access to all the phones OS and hardware features.

The great benefit with React Native as the core technology is that the code can be very similar to that of your React web application. This is because the React Native library is directly related to the web version. You have all the familiar concepts, like the usage of JSX, lifecycle events, Component class, render method, usage of modules like Redux and many other React capabilities.

## Chapter 20: Native Mobile Application development with React Native

The one big difference you will run into is that you have no HTML elements available in React Native. No `<div>`, `<span>`, `<ul>` etc. There is also a different approach to CSS for styling. All of this is kind of a pain, because you would have gone through a lot of work to get all of that written up in your React Web application. It does not turn out to be too much of a hassle though. Partly because you really should re-think your UI design when you are writing an application for a small mobile device screen. The website pages in many cases do not make sense to be rendered the same on a mobile device. Even if you are doing a responsive design with libraries like Bootstrap, that is still not good enough.

For now, the goal should be to take advantage of what you can from your React Web application. Then you can strive to write your mobile code once and be able to run the application on both iOS and Android. It is not always possible to write the code and run it on both iOS and Android, but you can come close if you work at it.

*Note: This book does not cover the topic of how to integrate low-level native code into your application. This is a technique where you actually write code in Objective-C for iOS or Java for Android. NewsWatcher did not require this to be done, and you would need to understand when and how you would need this advanced technique.*

# 20.1 React Native starter application

The React Native application requires a certain amount of scaffolding to be in place in order to be able to be built and run. As one option, you could copy a project from someone else from GitHub. There are also tools to generate a starter application. The second choice is what I went for.

What I did was to run a command to get my React Native application created and then I brought over code from the web application version. I was able to prove how easy it is to take the React website code and write a native application from that. Certainly, most of the code logic is going to be the same. It is mostly the UI markup that would have to change. I will now go through all the steps I took to get the React Native starter application up and running.

**Step One: Install a command line utility tool and a Desktop Development Tool**
With Node installed, you could use NPM to install the create-react-native-app command line utility. This is used to create a starting point application. Another option would be to use the Expo npm utility that you can also install from npm. I will use Expo. Expo is fully supported and even mentioned on the official React Native site.

# PART III: The Presentation Layer (React/HTML)

Here is the command to get Expo installed:

```
npm install exp --global
```

The exp utility will be used to generate your starter application and eventually create the finished application file that will be uploaded to the app stores for others to install, such as in Google Play.

You will also need to install a desktop application that acts as a development environment for Expo. It is called Expo XDE (Expo Development Environment). To install XDE you can go to the expo website and download the version you need. There are downloads depending on the OS of your machine.

### Step Two: Create the starter application
Now you can create the React Native application. To do that, you will use the Expo utility you just installed. Open a command prompt and create a folder where you want the application to be created in. Then run the following command, replacing the text with what you want your application called:

```
exp init my-new-project
```

Alternatively, you can launch the XDE desktop application and from a menu selection, create an application. This is actually what I did. I selected the template that creates an application with some navigation UI already in place.

*Note: When you run XDE you will need to provide a username and password. This is because XDE requires an account to be used for the generation of the necessary file for eventual uploading to an app store.*

### Step Three: Run the application on your mobile phone
With the XDE application open, you can select to have the application you build to be run on your phone. Before doing that, you need to first configure your phone for tethering and actually have it connected via USB to your computer.

To set up tethering and debugging on an Android phone you can do the following:

Go to **Settings** and under **General** you can scroll until you find the selection for **About phone**. Open that and then tap on **Software info**. Tap **Build number** (you may have to tap it multiple times) and then you will see a message that you have enabled developer mode.

Now go to the **General** settings to the **Developer options** selection. Select **USB debugging** to turn it on.

Chapter 20: Native Mobile Application development with React Native

You are all set to make changes to some code and experiment and see what you can do with React Native. I found that with my Windows machine, there were times that the connection to my phone did not seem to work too well. For me, the "localhost" selection, from the settings worked the best as the protocol.

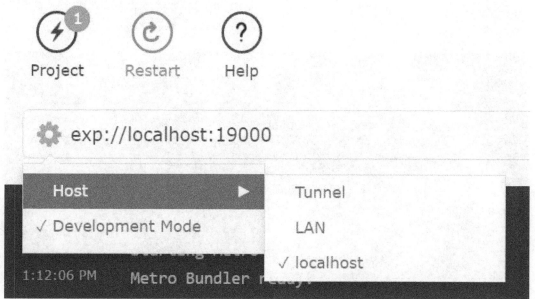

*Figure 108 - Setting the connection protocol*

Now, you can go to the XDE UI and open your project and then you will be able to click on the **Device** button and select **Open on Android**. The application will load and be started on your phone and you can interact with it.

If you recall, there were command line scripts to put together the application bundle for deployment. With the usage of Expo XDE, all of that happens as you open a project and click to have it run on your phone.

*Note: You can start up a phone emulator on your PC that can run the application in it. This allows you to do this in place of using your physical phone that is connected via USB. You can create virtual devices for many different types of phones, including iPhones. To do that you need to install something like Xcode, Android Studio, or the Genymotion emulator.*

The following screen shows an entry for the tethered phone that you would select to have your application run on. If you had an emulator running, the emulator would also show up. You can also set up configuration to have the application auto-load any code changes that are being made in the editor. This is called hot reloading.

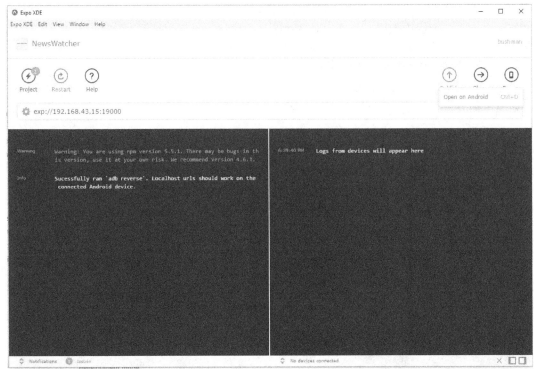
*Figure 109 - Expo Application*

## 20.2 Components

React Native has a set of components you can use for doing things like displaying text, getting input and displaying images. Besides this, there are plenty of UI component libraries that you can install from NPM that can give you what you are looking for. For example, if you like Material-UI from Google, for your look and feel, there are libraries that implement that for React Native.

You can now learn about a few of the simple components available and see that they are used like any other component that you have seen. You use them in a render function and pass needed props to them in typical React code you have seen before.

The `View` and the `Text` components are some easy components to start your learning with. `View` is basically what a `div` was in HTML. It is a container for other components to help with the UI layout.

# Chapter 20: Native Mobile Application development with React Native

The `Text` component is for displaying text. The following code is close to what you get when you create a React Native Application from Expo. You can find a file names App.js that is the starting point for rendering your UI.

```
import React from 'react';
import { Text, View } from 'react-native';

export default class App extends React.Component {
  render() {
    return (
      <View style={{borderTopWidth: 30}}>
        <Text>Open up App.js to start working on your app!</Text>
      </View>
    );
  }
}
```

This code should mostly look familiar, as you have the import statements and get the React usage from that, as well as the `View` and `Text` components. The `App` uses the Component class and implements a render function that you see returning the `View` with the `Text` inside.

*Figure 110 – Simple starter application*

There are around three dozen components to use to construct your application. These can be found in the online documentation for React Native. This section will only cover just a few of them. You will see more of the usage of these in the NewsWatcher code.

357

There is a `TextInput` component that is used when you want to collect text from a user. It has a prop called onChangeText that you provide a callback function to, for capturing the text that is entered.

There is a `Button` component and it is exactly what you would think it is. It has an onPress prop that you provide a callback function for. This will run on the button press.

There are other components for touch interaction, such as `TouchableHighlight`, `TouchableOpacity` and `TouchableElement`. For example, you might have an image that can be touched, or a bunch of items in a list that each have their own touch response. The `onPress` prop is used to provide a callback that can be used to process the touch event.

There are several components you can use to display lists of information. If you have a small set of data to render in a list, you can use the `ScrollView` component. You will have a performance problem if the list has a lot of elements.

There is a component named `FlatList` that only renders the data that is in view as the user scrolls. You can also use the `pagingEnabled` prop for paging through data and fetching it on demand in batches as scrolling happens.

There are many more components to investigate, such as for displaying images, radio buttons, checkboxes, selection picking and much more. Read the official documentation to learn about all that is available (https://facebook.github.io/react-native/docs/components-and-apis.html).

*Note: It may soon be possible to have a common component library that can even be used for your Web Site React code and your React Native application. You will most likely still need to make some modifications to the UI design on the mobile application because of the screen size.*

# 20.3 Styling your application

As was mentioned, HTML elements are not available to use in a React Native application. CSS is also not available. Instead of using CSS, there is a special way of applying style right in the JavaScript you write. To do the styling, each core components of React Native all accept a prop named style. This is close to how CSS works on the web, however, names are written using camel casing. For example, you use `backgroundColor` instead of `background-color`.

While you can use a JavaScript object right inside of the prop, it is much cleaner to split this out and use the `StyleSheet.create` function to define several styles in one place. You

## Chapter 20: Native Mobile Application development with React Native

then reference that in your render function code. Here is an example that shows the embedded object and the usage of the function mentioned:

```
import React from 'react';
import { StyleSheet, Text, View } from 'react-native';

export default class App extends React.Component {
  render() {
    return (
      <View>
        <View style={{width: 50, height: 50, backgroundColor: 'blue'}} />
        <Text style={styles.myTextStyle}>This is some text to experiment with</Text>
      </View>
    );
  }
}

const styles = StyleSheet.create({
  myTextStyle: {
    color: 'blue',
    fontWeight: 'bold',
    fontSize: 12,
  },
});
```

Width and height are set with properties and use values that will cause the sizing to be the same across devices, regardless of resolution. There is a `View` component used to draw a blue rectangle. The rectangle area is 50 by 50. The Text component brings in the style from the returned object of the `StyleSheet.create()` call. The rectangle and the text are both blue. The following image shows how this will look.

*Figure 111 - Styling*

# 20.4 Layout with flexbox

Flexbox is a standard that exists for web page layout. You can read the Mozilla (MDN) documentation to learn all the fine details. React Native takes this standard and makes use of it within its styling capability. What this adds, are properties you can add to your styling object that specify exact layout areas.

359

# PART III: The Presentation Layer (React/HTML)

The first property to know about is the `flex` property. This is what you can use to specify relative sizing of anything. For example, you can have three rectangles and give them each a flex value of 1 and then relative to each other, they will get sized. Since they each have a value of one, they each take up the same amount of space.

This layout with flexbox can be hierarchical, meaning that a parent can be part of a flex layout at its level, and then inside of that there, is another flex layout that is applied.

By default, the flex layout is determining the vertical layout of the area (called column layout). You can alter that and make it specify the horizontal layout (called row layout). You can also specify the justification, alignment and spacing.

Here is some sample code that has two subsections. One has a `flex` value of two and the other a value of one. This means that the top area will be twice the size of the lower area. Study the following code:

```
import React from 'react';
import { StyleSheet, Text, View } from 'react-native';

export default class App extends React.Component {
  render() {
    return (
      <View style={styles.container}>
        <View style={styles.subContainer1}>
          <Text style={styles.childText}>One</Text>
          <Text style={styles.childText}>Two</Text>
          <Text style={styles.childText}>Three</Text>
        </View>
        <View style={styles.subContainer2}>
          <Text style={styles.childText}>Four</Text>
          <Text style={styles.childText}>Five</Text>
          <Text style={styles.childText}>Six</Text>
        </View>
      </View>
    );
  }
}

const styles = StyleSheet.create({
  container: {
    flex: 1,
    marginTop: 25
  },
  subContainer1: {
    flex: 2,
    flexDirection: 'row'
  },
```

# Chapter 20: Native Mobile Application development with React Native

```
  subContainer2: {
    flex: 1
  },
  childText: {
    flex: 1,
    borderWidth: 2,
    textAlign: 'center',
    fontSize: 16
  },
});
```

You also see that one of the flexbox areas has a `flexDirection` set to 'row'. This makes the components inside of it be spread out horizontally. Here is what the result looks like:

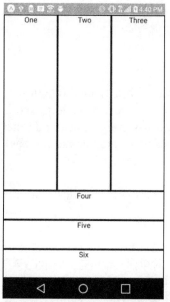

*Figure 112 - Flexbox layout hierarchy*

There are many more flexbox styling properties you can take advantage of. The main point of flexbox is to get you away from specifying pixel dimensions for anything and having everything sized relative to each other. That way no matter the device, it can be displayed properly.

361

## 20.5 Screen navigation

React Native does not have a built-in capability to set up any kind of screen transitions. Instead, there is a very popular npm library that you can install to handle this. You can do an npm install of react-navigation and then use that to set up different types of navigation. For example, if you want to treat your navigation like a stack, where you can go from one screen to another and as you transition, the previous screens stay in the stack history that you can go back to. Here is a simple example that shows having a screen with a clock on it and another screen that shows the weather on it.

```
import {StackNavigator} from 'react-navigation';
const App = StackNavigator({
  Clock: { screen: ClockScreen },
  Weather: { screen: WeatherScreen },
});
```

The `ClockScreen` and `WeatherScreen` are just components with a render function to display whatever you can imagine for each of those screens. The important code to see is simply the setting up of the use of the Stack Navigator. This allows you to have code that calls a navigate function to go to either of the screens and even pass in parameters. The back button on an Android device will also take you back a screen in the stack.

The header provided by `StackNavigator` will contain a back button to go back from the active screen if there is one.

The other mode of navigation is that of using tabs. You see this type of navigation all the time in mobile applications. You will see the tabs at the top or the bottom. The code is exactly the same, but instead of using `StackNavigator`, you use `TabNavigator`. Once you do so, you will see the tab appear for you to make your screen selection. You can add styling to this to add icons.

The navigation can be set up to be nested. For example, you can have the main navigation with tabs, but within each screen for the tab, you can set up some stack navigation, such as having a tab with a list, and clicking on an item in the list doing a stack navigation to a details screen.

## 20.6 Device capability access

The React Native API is still growing in its capabilities. This is true as far as being able to access the capabilities of the phone hardware. For example, you can access the phone's camera roll of photos, but not the camera to take a photo. You can get at the geolocation, but not at the accelerometer. It can be assumed, that over time, most phone capabilities will be exposed through the React Native API.

Until that time, you can simply use the API provided by Expo. Besides providing the UI to create and run applications, there has been provided a library in npm that you can use in your code to access thing like the camera, accelerometer, fingerprint, gyroscope, contacts and much more.

To use the Camera component, you use it like other React components. This means that you will need to render it as an element first in some JSX. To actually take a photo, you need to get a reference to the component and then call a function on it. Here is what it would look like.

```
// in the render() function
<Camera ref={ref => { this.camera = ref; }} />

// Somewhere in code, perhaps in response to a touch event
this.camera.takePictureAsync().then(data => console.log("Took a photo"));
```

## 20.7 Code changes to NewsWatcher

You can now learn about the code changes that are needed to get the NewsWatcher application written as a native application.

### The starting point

As with our other React code, this code also has an App.js file that acts as the starting point for your logic to render and run all your code.

The code will set up the navigation, including the UI for the navigation bar and all the individual screen components that will make up the application. The navigation bar is dependant on the mobile platform, because iOS and Android UI navigation can look different. Thus, there is code that determines which to render. The Platform.os property holds a string that will tell you which mobile platform you are running on.

# PART III: The Presentation Layer (React/HTML)

You will see Redux being used in the eact same way as before. Here is all of the code for the start up.

```
// App.js
import React from 'react';
import {Platform,StatusBar,StyleSheet,View,AsyncStorage} from 'react-native';
import { AppLoading, Asset, Font } from 'expo';
import { Ionicons } from '@expo/vector-icons';
import { TabNavigator } from 'react-navigation';
import MainTabNavigator from './navigation/MainTabNavigator';
import { createStore } from 'redux';
import { Provider } from 'react-redux'
import reducer from './reducers'
import { fetchMyNews } from './utils/utils';
import { fetchMyProfile } from './utils/utils';

const store = createStore(reducer);
const RootNavigator = TabNavigator(
  {
    Main: {
      screen: MainTabNavigator,
    },
  },
  {
    navigationOptions: () => ({
      headerTitleStyle: {
        fontWeight: 'normal',
      },
    }),
  }
);

export default class App extends React.Component {
  state = {isLoadingComplete: false };
  componentDidMount() {
    // Check for token in device local storage, meaning user is signed in
    AsyncStorage.getItem('userToken', function (err, value) {
      if (value) {
        const tokenObject = JSON.parse(value);
        store.dispatch({ type: 'RECEIVE_TOKEN_SUCCESS', msg: `Signed in as ${tokenObject.displayName}`, session: tokenObject });
        fetchMyNews(store.dispatch, tokenObject.userId, tokenObject.token);
        fetchMyProfile(store.dispatch, tokenObject.userId, tokenObject.token);

      }
    })
  }
```

## Chapter 20: Native Mobile Application development with React Native

```
  render() {
    if (!this.state.isLoadingComplete && !this.props.skipLoadingScreen) {
      return (
        <AppLoading
          startAsync={this._loadResourcesAsync}
          onError={this._handleLoadingError}
          onFinish={this._handleFinishLoading}
        />
      );
    } else {
      return (
        <Provider store={store}>
          <View style={styles.container}>
            {Platform.OS === 'ios' && <StatusBar barStyle="default" />}
            {Platform.OS === 'android' && <View style={styles.statusBarUnderlay} />}
            <RootNavigator />
          </View>
        </Provider>
      );
    }
  }

  _loadResourcesAsync = async () => {
    return Promise.all([
      Asset.loadAsync([
        require('./assets/images/robot-dev.png'),
        require('./assets/images/robot-prod.png'),
      ]),
      Font.loadAsync({
        // This is the font that we are using for our tab bar
        ...Ionicons.font,
        'space-mono': require('./assets/fonts/SpaceMono-Regular.ttf'),
      }),
    ]);
  };

  _handleLoadingError = error => {
    // In this case, you might want to report the error to your error
    // reporting service, for example Sentry
    console.warn(error);
  };

  _handleFinishLoading = () => {
    this.setState({ isLoadingComplete: true });
  };
}

const styles = StyleSheet.create({
  container: {
    flex: 1,
    backgroundColor: '#fff',
  },
```

```
    statusBarUnderlay: {
      height: 24,
      backgroundColor: 'rgba(0,0,0,0.2)',
    },
});
```

# Navigation UI

The react-navigation module is used that requires an object as part of its construction. There are different types of navigations available, such as stack and tab. We will be using tab by using the `TabNavigator` function. You pass in an object to that and specify the files that hold the screen UIs and also the tabs and icons for the toolbar to use for each.

Here is the code for the configuring of the navigation for the NewsWatcher application:

```
// MainTabNavigator.js
import React from 'react';
import { Platform } from 'react-native';
import { Ionicons } from '@expo/vector-icons';
import { TabNavigator, TabBarBottom } from 'react-navigation';
import Colors from '../constants/Colors';
import HomeScreen from '../screens/HomeScreen';
import MyNewsScreen from '../screens/MyNewsScreen';
import LoginScreen from '../screens/LoginScreen';
import ProfileScreen from '../screens/ProfileScreen';

export default TabNavigator(
  {
    HomeNews: {
      screen: HomeScreen,
    },
    MyNews: {
      screen: MyNewsScreen,
    },
    NewsFilters: {
      screen: ProfileScreen,
    },
    Account: {
      screen: LoginScreen,
    },
  },
  {
    navigationOptions: ({ navigation }) => ({
      tabBarIcon: ({ focused }) => {
        const { routeName } = navigation.state;
        let iconName;
        switch (routeName) {
          case 'HomeNews':
            iconName =
              Platform.OS === 'ios' ?
                `ios-home${focused ? '' : '-outline'}`
```

## Chapter 20: Native Mobile Application development with React Native

```
                    : 'md-home';
              break;
            case 'MyNews':
              iconName =
                Platform.OS === 'ios' ?
                  `ios-funnel${focused ? '' : '-outline'}`
                  : 'md-funnel';
              break;
            case 'NewsFilters':
              iconName =
                Platform.OS === 'ios' ?
                  `ios-options${focused ? '' : '-outline'}`
                  : 'md-options';
              break;
            case 'Account':
              iconName =
                Platform.OS === 'ios' ?
                  `ios-person${focused ? '' : '-outline'}`
                  : 'md-person';
          }
          return (
            <Ionicons
              name={iconName}
              size={28}
              style={{ marginBottom: -3 }}
              color={focused ? Colors.tabIconSelected :
Colors.tabIconDefault}
            />
          );
        },
      }),
      tabBarComponent: TabBarBottom,
      tabBarPosition: 'bottom',
      animationEnabled: false,
      swipeEnabled: false,
    }
  );
```

The iconName property contains a value specifying the vector graphic icon to use. These need to be set for values depending on iOS and Android. There is a website to go to for seeing the list of possible icons. You can search the site https://expo.github.io/vector-icons/.

Now you can consider each of the screens that are simply React Component derived classes that React Native navigation uses to render as they are brought into usage.

## The Home News screen

This is code for showing the home page screen that contains the default top news stories. You will see much that is familiar. You see that there is still a `componentDidMount()` function that contains the backend fetch call. The local component state is used for that storage.

367

# PART III: The Presentation Layer (React/HTML)

In the scrollable list of news stories, you see something new. This is the `TouchableNativeFeedback` React Native component. This is used as the outer parent for all of what is contained inside and gives the ability to capture a touch that will launch the browser to display the news story. Everything else is a fairly straightforward translation from the HTML elements to the ones that are supported by React Native.

```
// HomeScreen.js
import React from 'react';
import {
  Image, Platform, StyleSheet,
  Text, TouchableHighlight,
  TouchableNativeFeedback, View,
  ScrollView, Alert
} from 'react-native';
import { WebBrowser } from 'expo';
import { toHours } from '../utils/utils';

export default class HomeScreen extends React.Component {
  constructor(props) {
    super(props);
    this.state = { isLoading: true, news: null };
  }

  static navigationOptions = {
    title: 'HomeNews',
  };

  componentDidMount() {
    return fetch('https://www.newswatcher2rweb.com/api/homenews', {
      method: 'GET',
      cache: 'default'
    })
      .then(r => r.json().then(json => ({ok:r.ok, status:r.status, json})))
      .then(response => {
        if (!response.ok || response.status !== 200) {
          throw new Error(response.json.message);
        }
        for (var i = 0; i < response.json.length; i++) {
          response.json[i].hours = toHours(response.json[i].date);
        }
        this.setState({
          isLoading: false,
          news: response.json
        });
      })
      .catch(error => {
        this.props.dispatch({ type: 'MSG_DISPLAY',
            msg: `Home News fetch failed: ${error.message}` });
        Alert.alert(`Home News fetch failed: ${error.message}`);
      });
  }
```

## Chapter 20: Native Mobile Application development with React Native

```
onStoryPress = (story) => {
  WebBrowser.openBrowserAsync(story.link);
};

onNYTPress = () => {
  WebBrowser.openBrowserAsync('https://developer.nytimes.com');
};

render() {
  if (this.state.isLoading) {
    return (
      <View>
        <Text>Loading home page news...</Text>
      </View>
    );
  }

  let TouchableElement = TouchableHighlight;

  if (Platform.OS === 'android') {
    TouchableElement = TouchableNativeFeedback;
  }
  return (
    <View>
      <ScrollView>
        {this.state.news.map((newsStory, idx) =>
          <TouchableElement key={idx}
                     onPress={() => this.onStoryPress(newsStory)}>
            <View style={styles.row}>
              <View style={styles.imageContainer}>
                <Image source={{ uri: newsStory.imageUrl }}
                     style={styles.storyImage} />
              </View>
              <View style={styles.textContainer}>
                <Text style={styles.storyTitle} numberOfLines={2}>
                  {newsStory.title}
                </Text>
                <Text style={styles.storySnippet} numberOfLines={3}>
                  {newsStory.contentSnippet}
                </Text>
                <Text style={styles.storySourceHours}>
                  {newsStory.source} - {newsStory.hours}
                </Text>
              </View>
            </View>
          </TouchableElement>
        )}
        <TouchableElement key={this.state.news.length}
                     onPress={() => this.onNYTPress()}>
          <View style={styles.row}>
            <Image
              source={require('../assets/images/poweredby_nytimes_30b.png')}
            />
```

```
              <View style={styles.textContainer}>
                <Text style={styles.storyTitle} numberOfLines={2}>
                  Data provided by The New York Times
                </Text>
              </View>
            </View>
          </TouchableElement>
        </ScrollView>
      </View>
    )
  }
}

const styles = StyleSheet.create({
  row: {
    alignItems: 'center',
    backgroundColor: 'white',
    flexDirection: 'row',
    borderStyle: 'solid',
    borderBottomColor: '#dddddd',
    borderBottomWidth: StyleSheet.hairlineWidth,
    padding: 5,
  },
  imageContainer: {
    backgroundColor: '#dddddd',
    width: 90,
    height: 90,
    marginRight: 10
  },
  textContainer: {
    flex: 1,
  },
  storyImage: {
    width: 90,
    height: 90,
  },
  storyTitle: {
    flex: 1,
    fontSize: 16,
    fontWeight: '500',
  },
  storySnippet: {
    fontSize: 12,
    marginTop: 5,
    marginBottom: 5,
  },
  storySourceHours: {
    fontSize: 12,
    color: 'gray',
  },
});
```

## Chapter 20: Native Mobile Application development with React Native

Here is what that UI looks like on my phone fo the home news screen:

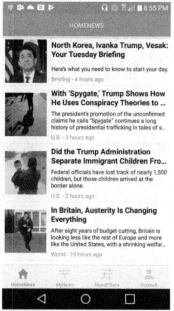

*Figure 113 – Home news screen*

If you are looking closely and comparing the code of the React Web app with that of the React Native app, you will notice that we don't actually use the `componentDidMount()` function in the React Native code in some places. You will see that the `ProfileScreen` and `MyNewsScreen` components don't have that function like the `HomeScreen` component does.

There are a few reasons why. One difference is that this gets called only once and it gets called even before you switch to each screen. For example, it gets called on the profile screen at the startup time of the app even though the profile screen is not being shown. The fetch needs to make use of the session token at that time to get data for the user. The problem is that there is a race condition with the App.js that gets the token from the local storage and it might not be available yet. The fix is to place the fetching when you know the session is available.

The `HomeScreen` component does not need a JWT token to call the endpoint API it is using, so there is no timing issue.

PART III: The Presentation Layer (React/HTML)

# The other screens

Similar code transformations happen for making the other screens. You can look at the code for the `LoginScreen`, `MyNewsScreen` and `ProfileScreen` components. I will give a few highlights and give you screen images to look at.

**Modal:** You will see the usage of the `Modal` component. In this case, it is a bit different from what we did in the React web code. Here, there is actually a visible prop that we set to control if it is visible or not.

**Buttons:** You will see in the code that I experimented with making buttons using both the `Button` and the `TouchableOpacity` components. The latter is simply more versatile, but both behave and look similar.

Here is what that UI looks like on my phone for a modal component and the buttons:

Figure 114 – Modal UI

**Keyboard:** The `KeyboardAvoidingView` component is used to wrap UI that will move out of the way when a keyboard interaction UI is moved into place. This is used around the `TextInput` components. The `TextInput` themselves also have some interesting props to use to do things like stating what type of keyboard to display and how to transition to another field.

Chapter 20: Native Mobile Application development with React Native

Here is what that UI looks like on my phone for the keyboard usage:

Figure 115 – Keyboard UI

**Picker:** The `Picker` component is used to display a type of drop-down for making selections. The selections, in this case, are the news filters that you can set up and choose between.

Here is what that UI looks like for the profile screen where the user specifies their news filters:

Figure 116 – New Filter profile UI

373

## Application configuration file

There is a special file that is provided that controls some aspects of the React Native application. You can open and look at the app.json file to view what it contains. You can see that you specify things like the splash screen image and also what the description is for the application as shown in the app store. One of the things I needed to do for the Android section was to specify the `WAKE_LOCK` permission. This is set to keep the screen from dimming and the processor from going to sleep.

```
// app.json
{
  "expo": {
    "name": "NewsWatcher",
    "description": "News serving application",
    "slug": "NewsWatcher",
    "privacy": "public",
    "sdkVersion": "23.0.0",
    "version": "1.0.0",
    "orientation": "portrait",
    "primaryColor": "#cccccc",
    "icon": "./assets/images/icon.png",
    "splash": {
      "image": "./assets/images/splash.png",
      "resizeMode": "contain",
      "backgroundColor": "#ffffff"
    },
    "ios": {
      "supportsTablet": true
    },
    "android": {
      "versionCode": 2,
      "package": "com.blueskyproductions.newswatcher2rweb",
      "permissions": ["WAKE_LOCK"]
    }
  }
}
```

# 20.8 Application Store Deployment

After the application is all developed, you will want to make it available in one of the application stores. I will walk through some of the details you need to know to get this up and on the Google Play store. What you will do is create the APK standalone binary that is used to distribute your application.

## Chapter 20: Native Mobile Application development with React Native

When using Expo, the standalone app knows to look for updates at your app's published URL. You can later run a command (exp publish) to upload a new version. The next time a user opens the app they will automatically download the new version. This is called an "Over the Air" (OTA) update and is built into Expo so you don't need to install anything.

To build the APK (or iOS IPA file), open a Node.js command prompt. A regular windows DOS prompt is not capable of running the commands. You need to run the following command in the directory of your project.

```
exp build:android
```

This process will transfer the application code up to a cloud server run by Expo and a build process will run there. You can see that it is progressing if you go to the link provided for you each time. Once a build is done, you can also install it directly from the Expo portal with the Expo app on your phone by snapping an image of the QR code you can view.

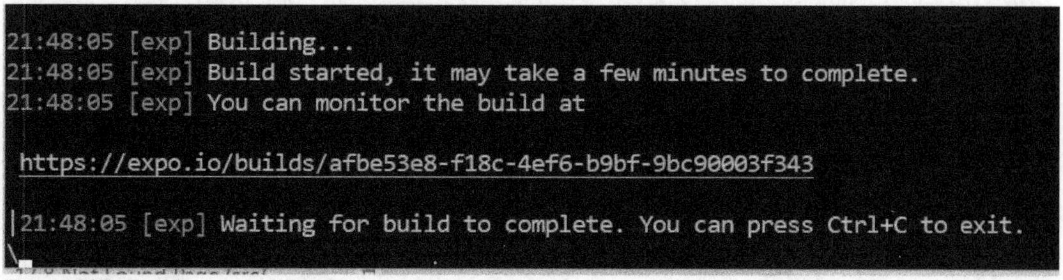

```
21:48:05 [exp] Building...
21:48:05 [exp] Build started, it may take a few minutes to complete.
21:48:05 [exp] You can monitor the build at

 https://expo.io/builds/afbe53e8-f18c-4ef6-b9bf-9bc90003f343

|21:48:05 [exp] Waiting for build to complete. You can press Ctrl+C to exit.
```
*Figure 117 - Build output*

It could take five or so minutes to complete. Once it is done, you are presented with a URL to use to download the APK file in the case of an Android build. You run a similar command to create a file for iOS.

Once you have downloaded the file, you can upload it to the app store. To get that up and available to Android users through the Google Play store you need to have an account to do so and go to the portal to work through that. Once your application is available in the store, you can see it at https://play.google.com/store and of course open the Play Store on your phone and install it.

To upload your application file for Android and publish it on Google Play you can go to https://play.google.com/apps/publish. You can click the button to create a new application. You can begin by publishing your app as an Alpha release. You can click on **Release management** on the left menu are and then choose **App releases**. Then on that screen, you will see where you can click to manage the Alpha release. You can create a release and upload your .APK file.

375

## PART III: The Presentation Layer (React/HTML)

You can continue to upload newer updates to your APK file as you add features and fix bugs. Alternatively, Expo can be used to publish releases that don't require you to go through the Google Play portal, but are updates that the application will automatically download on the mobile device.

Once you have the APK file uploaded to Google Play, you can click on the Alpha release and add testers to it. You can email a specific URL that will be available to any friends that would be able to help test the application. You can then promote it to Beta and finally to Production where it will be seen by everyone in the Google Play application store.

If you have an Android phone, you can open Google Play and search for "NewsWatcher" and install the application I uploaded there already.

*Note:* There are still a few kinks to work out to get Enzyme, with Jest to be able to do a mount and shallow rendering in test code with them. That would be a great research project on its own. There is documentation in Enzyme for this, but it still does not carry it deep enough to be able to do testing of React Native as it does with React for the web.

**Letting Google Play manage the public signing key**
There is one thing I omitted to make clear in the previous text. That is that if you upload the application with the Expo signed key, you can't ever upload another APK file, as Google Play will complain about a number of things. For example, it will complain that the same version number is being used already. But if you change that, then Expo generates a new key for it and Google Play will not accept it. Read the Expo documentation on this and they make it clear that you can opt into Google Play doing the user signed key and then you can set up Expo to let you manage signing with a provided upload key that you give it.

Bottom line, it seems like you would want to go with Google Play managing the signing so that you can upload new APK files. Otherwise, you are tied into using Expo always to do the updates in the app through their mechanisms. I know this all sounds confusing at first, so just try and follow along here.

To start with, you need to open Android Studio and just create a new application. Then once that is created, click on the **Build** menu and select **Generate Signed APK**. There will be a button there named **Create New**. Click that and fill in the form, such as follows:

# Chapter 20: Native Mobile Application development with React Native

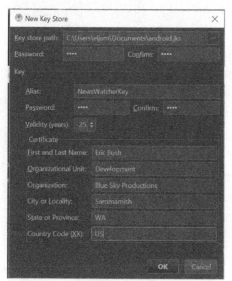

*Figure 118 – Creating an upload signing key*

Follow through and do the APK generation. You will not be needing the APK file, but will be taking the jks file and using that when you do the Expo build. Make sure to take note of the passwords and the alias you typed in. Now, when you run the Expo build command and select to provide your own key, you will be asked for the path and name of the jks file. Follow the prompts and you will get an APK file you can use with Google Play now. You will use this file over and over to do the signing every time you want to upload a new version of your application to Google Play.

**Note:** *I utilize Expo in this book because it is the easiest and fastest way to get something up and working. You may decide that you want the full flexibility to create and manage your React Native application yourself. In that case, you should create the application with create-react-native-app installed with NPM. You install this and can do everything you need from the command line, or even with the use of Android Studio.*

# CONCLUSION

A lot of material has now been covered and you should at this time feel really good about your newly acquired knowledge and skills. You started by learning about what a three-tier architecture is and how a MERN JavaScript full-stack implementation is a great choice for an implementation.

The NewsWatcher sample application was featured throughout the book as a full end-to-end implementation of the architecture you have been learning about. You now have the code that you can refer to and base any of your own work on.

The three parts of the book covered the layers of a three-layer architecture and gave you foundational knowledge about each layer. You also learned about the SDKs to use and specific usage scenarios of everything for the NewsWatcher application.

You can now go forward and do your own learning for your own specific needs and be successful. You may choose to just work in one of the layers, or you may be able to contribute across all layers. It is always good to be knowledgeable overall as you will know better how everything works internally and realize how what you do in one layer affects another layer.

# INDEX

## A

Access token · 153
aggregation · 145
alerting · 89
APK · 374
APM · 258
Atlas portal · 77
atomic transaction · 61
authentication · 154

## B

Bootstrap · 291
BSON · 24

## C

callback · 107
Chrome Developer tools · 273, 344
CI/CD · 251
CloudWatch · 256
collection · 22
Components · 279
concurrency · 147
Content Security Policy · 159
create · 141
create-react-app · 268
Criteria operators · 46
CRUD · 41, 58, 141
CRUD operations · 141
CSRF · 158
CSRF - Cross-Site Request Forgery · 158

## D

DAL · 16
data backup · 91
data layer · 17
Data Layer · 5
data model · 29
database index · 62
DBMS · 15
DDoS · 156
debugging · 227, 344
delete · 144
denormalize · 30
development stack · 4
DevOps · 88, 244
disaster recovery · 91
DNS · 221
document · 23
document types · 39
document-based · 25
Document-based · 19
document-based database · 19
documents · 27
DoS · 156
DOS/DDOS attack · 156

## E

Elastic Beanstalk · 171, 180
Enzyme · 338
Expo · 353
Express · 115
Express middleware · 122, 124, 129
Express Middleware · 122
Express routing · 118
Express static · 128
Express template · 137
Express.js · 5

## F

Fiddler · 229
findOneAndUpdate · 148
Flexbox · 359

## G

gateway layering · 97
Google Play store · 374

## H

Helmet · 159
HTML · 266
HTML forms · 286
HTTP status code · 134
HTTP verbs · 118
HTTP/Rest requests · 293
HTTPS · 154, 220

## I

IaaS · 6
index sort order · 66
indexing · 64, 66, 68, 85

## J

JavaScript · 6, 100, 102
Jest · 339
JSON · 25
JSON Web Token · 153
JSX · 277
JWT · 153, 197

## L

lifecycle hooks · 287
lint · 242
load balancing · 170

load testing · 241
logging · 244

## M

memory leak detection · 249
metrics · 89
Microservices · 173
middleware · 186
middleware error handling · 127
middleware errors · 127
mLab · 20
mocha · 235
module design · 110
modules · 103, 113
MongoDB · 5, 19, 20, 22, 23, 29, 41, 188
MongoDB Atlas · 77
MongoDB Compass · 79
mongodb module · 139, 140
MongoDB replica set · 69
monitoring · 88, 256
mount · 341
MVC · 282

## N

NewsWatcher · 8, 76, 81, 184, 302
nextTick · 152
Node · 101
Node Package Manager · 110
Node.js · 5, 98, 100
Node.js framework · 107
Node.js internals · 160
Node.js modules · 113
normalization · 31
normalized · 32
NPM · 110
npm install · 105
NPM modules · 104

## O

Object Data Mapping · 146
Object Relational Mappings · 146

Optimistic concurrency · 149

## P

PaaS · 6
Polymorphic Documents · 56
presentation layer · 263
Presentation Layer · 5
process forking · 187, 212
processing loop · 163
profiling · 245
projection criteria · 53

## Q

queries · 41, 85
query criteria · 44, 45
query operators · 46

## R

React · 5, 265
React Native · 352
React Router · 289
React state · 281
react-redux · 298
read · 142
reducer · 297
Redux · 294
referencing · 32, 35
render() method · 276
REPL · 101
replica set · 40, 69
replication · 70
request · 117
request object · 132
response object · 134
Response object · 134
REST API · 152
REST Web Service API · 192
RESTful · 152
Route paths · 119
router · 126
router object · 126

## S

SaaS · 7
scaling · 168
security · 153
selenium · 335
SEO · 348
Server Side Rendering · 347
server.js · 185
service attacks · 155
service layer · 93, 95, 96, 98
Service Layer · 5
Service Oriented Architecture · 170
setInterval · 151
setTimeout · 103
shallow · 341
sharding · 40, 70
sharding key · 74
single thread · 106
SOA · 5
SPA · 263
SSR · 347
StackNavigator · 362
storage capacity scaling · 40

## T

thread pool · 164
three-tier · 5
three-tier architecture · 95
TLS/SSL · 154
topology · 12

## U

UI automation testing · 335
update · 143
update operators · 58
URL path · 121

## V

V8 · 162
Visual Studio Code · 7

381

VS Code · 7

## X

X-Ray · 258
XSS · 157
XSS - Cross-Site Scripting · 157

CPSIA information can be obtained
at www.ICGtesting.com
Printed in the USA
BVHW01s2158080818
524004BV00008B/72/P